Renewable Energy
facts and fantasies

Book design by Charles Allen Harris.

For more information, contact 2GreenEnergy at
www.2GreenEnergy.com.

Published by Clean Energy Press.
ISBN: 0615388353
EAN-13: 9780615388359

Renewable Energy
facts and fantasies
the tough realities
as revealed in interviews with
25 subject matter experts

by Craig Shields
editor, 2GreenEnergy.com

CONTENTS

ACKNOWLEDGEMENTS

I couldn't possibly have completed this project without the love and support of my wife Becky and her eagle-eye proofreading, nor without the guidance and wisdom of my fine friends George Alger, Cameron Atwood, and Terry Ribb. I'd also like to mention the inspiration that I receive daily from my two beautiful children, Jake and Valerie, and from my parents, who, from the time I was a toddler, never stopped reminding me that I can do anything, as long as I put my mind to it. Though I'm not at all sure that's true, I will never cease to appreciate the thought. Thank you.

FOREWORD

The moral compass that lives deep within all of us commands various things of our behavior—most of which pertain to the present, such as prohibitions on stealing, cheating, lying, and so forth. But I believe that the majority of us feel some sort of ethical obligation toward future generations as well.

And hasn't this argument heated up considerably in the past few decades? When I was a young boy, we had the occasional public service announcement on our black and white television sets regarding littering, yet there were very few reminders that we should consider the quality of the future that we were leaving for subsequent generations.

So much has changed regarding our perception of the future—including the fact that modern communications media have brought it so much more "in our face" than ever before. We worry about rogue states with access to nuclear weapons, global climate change, the long-term implications of nuclear accidents like Chernobyl, economic collapse, runaway population growth, epidemic diseases, scarcity of food and water supplies—and the implications of all these things for the decades—or millennia—that lie ahead. As the years roll forward, we seem to be making more decisions that affect not only our own health and safety, but also that of our children and generations to come.

Making this discussion even more interesting is the vast polarization that these subjects seem to have engendered. For some reason that I have trouble understanding, the whole issue of sustainability seems to have driven a wedge deep into our culture. On one side are those who believe that the issues I've named above are real, that they pose imminent danger to society, and that we have a moral obligation to do something about them. On the other side are those who see no real imperative to change our lifestyle to deal with these problems—or who deny that they exist at all.

I have to think that those in the prior camp will find this book more interesting than those in the latter. But I want to ensure the reader—in whichever group he finds himself—that I have made no attempt to frame this discussion according to any political ideology. In fact, though I personally believe that renewable energy needs to become a critically important part of our culture, I deliberately chose the book's title to remind the reader that not everything is possible. We live in a world of tough realities—technological, economic, and political—and we all would do well to understand those realities if we are to have relevant discussions as to what we must do as a civilization.

My position is that renewables—in all forms—will fail without these three things moved into place. Great technology by itself is a sure loser. The world is littered with great ideas sitting on shelves—whether they may be breakthrough patents bought by competitors whose intent was to leave them fallow, or, more commonly, good but under-funded ideas.

But what about ideas that are well financed—but that happen to cut across the vested interests that may have a disproportionate amount of influence? To me, the fact that the oil industry employs seven lobbyists for each senator and representative in Washington speaks for itself. Do you think they'd spend hundreds of millions of dollars if that fortune weren't buying them something of far greater value?

OK, but what happens when a lot of money and political power come

together to support a truly terrible technology? Sadly, this happens more than you would like to think. Corn ethanol is an excellent example. Fortunately, things like this are short-lived. At a certain point, people figure out that they've been ripped off, and the outcry is so loud that eventually we move on to the next idea.

The point is that we need to consider each of these three points on the renewable energy "triumvirate" if we are to have intelligent, responsible discussions on the subject of renewables—and isn't that the reason you're reading this book?

—Craig Shields, 2010

INTRODUCTION

In the considerable amount of time I spend reviewing and responding to the comments on the blog at 2GreenEnergy.com, I must say that certain things continue to surprise me. In particular, I have to say that I'm shocked by the vitriol surrounding global warming. After all, isn't the argument moot? Don't we need to curtail our consumption of fossil fuels for a great number of other completely independent reasons?

For a moment, take the viewpoint that global warming is a hoax—or at least that addressing it is a quixotic pursuit. OK, but do you doubt that gas and diesel emissions cause cancer? That enriching certain governments in the Middle East endangers everyone of us? That the scarcity brought about by a declining supply of oil causes military conflict and unnecessary loss of human life? That the damage to fragile ocean ecosystems, which are becoming more acidified each year, is having a profound effect on food supplies and the larger biosphere?

It is for all these reasons that the world's attention seems to be so riveted on renewable energy. It's really very difficult to pick up a copy of the Wall Street Journal or the New York Times and not see half a dozen articles that are directly related to the imperative to create a sustainable way in which we, as a civilization, generate and consume energy. Very few people doubt that our current fixation on fossil fuels will implode on itself as our population grows and its hunger for energy increases over the coming few decades.

I offer here a survey of the subject of renewable energy, provided from a number of different perspectives that I hope will be useful to people trying to make sense of a subject of such enormous complexity. In fact, I find that the most interesting aspect of this whole topic is, in fact, its very complexity—the way it taps into virtually all our disciplines of thought and learning. Here, one finds pieces of math, physics, chemistry, biology, economics, sociology, marketing, psychology, international relations, history, and political theory. With the possible exception of Gaelic folk tunes, the entirety of what one can study in college all comes into play here.

The book's title is meant to serve as a reminder that not everything is possible. We live in a world in which we feel entitled to abundant and inexpensive energy—and there is no doubt that the migration to renewables comes at a short-term cost. In particular, some of the technologies by which clean energy is attainable at scale today are in a kind of nascent state, and would therefore be quite pricey if implemented by the gigawatt in the level of development in which we find them today. In fact, some of the ideas that represent the greatest promise (e.g., geothermal and concentrating solar power) lag many decades behind others that have gotten a head start (e.g., wind and photovoltaics).

We live in a world of tough realities—technological, economic, and political—and we all would do well to understand those realities if we are to have relevant discussions as to what we must do as a civilization.

In writing this book, I took what I believe to be excellent advice of numerous friends, and decided to base the entire project on interviews with the widest possible variety of subject matter specialists. As my readers at 2GreenEnergy.com can attest, I honestly try—at a surface level at least—to cover the gamut of the sciences, the business issues, and the underlying political foundation that collectively determine our best course vis-à-vis clean energy. But I think the reader has been very well served here in allowing me to confine my contribution to asking the questions, and relying on other more focused minds for the answers.

As far as organization is concerned, I have tried to lay this out in an intuitive format that flows well—one in which a given subject progresses naturally into the next; I can only hope that the reader will agree. So let us begin with some of the big issues at the core of the subject: the imperative. Exactly why must we migrate away from fossil fuels in the direction of alternative sources of energy? What are the basic truths about our energy habits on this planet—and where those habits are taking us?

We start with my interview with Matt Simmons—probably the world's best-known expert on Peak Oil—the concept that world production of oil has hit its maximum in terms of barrels per day—and that this fact has specific and potentially dire consequences.

From there we move to conversations with people who lay out the case for renewables from a number of different and interesting perspectives—and then to a series of discussions around the various clean energy technologies that appear to hold the most promise. The book concludes with a few conversations surrounding alternative fuel transportation, intelligent business management, economics, sustainable living—and other related topics.

In the appendix, the reader will find a short piece that I wrote on basic physics. I observe that, luckily for us all, the science behind renewable energy is, at its core, fairly simple and straightforward stuff. Though the nuances may get into more advanced math and physics, we learned the basic principles in high school. Those wishing a quick review of these ideas will find them laid out in that section.

I want to point out something that readers may find disconcerting: many of the people I've interviewed disagree with one another—not to mention with me. If this book has value, that value lies precisely in the fact that I did not fashion it around my own or any other single person's ideas on the subject, emphasizing certain viewpoints while censoring out others that did not align with what I may have wanted to present. Having said that, I offer in the conclusion, for what they're worth, a few of the personal judgements that I take away from this exercise.

Don't expect a surprise ending; there is none. We actually do live in a world of tough realities. No one can snap his fingers and turn on a safe, readily available, and cost-effective stream of renewable energy. We're headed in that direction if the right factors, (discussed at length here) all come together, but we're not there now. Perhaps the drama lies in the race: will we get there in time?

I hope that readers will feel free to make comments and suggestions on the website, 2GreenEnergy.com.

TOUGH REALITIES

THE POLITICAL SCENE

PEAK OIL

Matthew Simmons served as energy adviser to U.S. President George W. Bush. His book Twilight in the Desert: The Coming Saudi Oil Shock and the World Economy *is a thoughtful examination of oil reserve decline rates—a phenomenon that points to the idea of "Peak Oil"—the concept that the world's supply of effectively extractable oil is declining.*

Craig Shields: I read your book—Twilight in the Desert. It certainly explains your position very well, or at least the position that you had at that point. Excellent stuff.

Matt Simmons: Thanks very much.

CS: Peak oil, in my estimation, is one of probably four or five good reasons to head in this direction—and it's a heck of a good one.

MS: Oh, it's the only one that basically forces you to do it.

CS: Well, from the position of someone who might not know an awful lot about peak oil, or may even be skeptical of the position, how would you articulate it?

MS: The first thing is to make people understand what you mean by "peak oil." Unfortunately, most of the critics of peak oil think it means we've run out of oil, and it doesn't mean anything like that. It means we've peaked. And we're never really going to run out of oil. We'll always have some oil. We might not have any we want to use, but we'll always have oil. But when the flow rate starts to decline, you have to use less.

I'm going to stick on the supply side for a minute. What's unfortunate is how little data we have of high quality that really pins down how real this issue is. With the exception of a handful of key oil-producing countries who have created genuine petroleum transparency field by field, and that includes all the participants in the North Sea, and Mexico, we have no data on the flow-rate from the rest of the world's oilfields. Now, you can really dig around, which is what I spent a decade doing, and getting lots of data on the flow-rates per field, or you could wait until the producing countries' flow-rates have been in decline for three or four years before you say "well obviously that country has peaked in its oil supply."

CS: Right. But, what is the imperative of these people to tell you anything about it, and least of all, the truth?

MS: Well, it's in everyone's best interests to be honest about this because the people it hurts the worst are the producing countries. In fact, ironically, I had a reporter in Dubai from Bloomberg email me this morning to get some comments on an article she's doing on the 50th anniversary of OPEC. And her questions were, "Do you think OPEC will be viable 50 years from now?" And "What should they be doing to strengthen their organization," and "What changes do you think they should make?" I basically said that I think within 20 years most of the OPEC producers will not be able to export oil anymore, because their internal use is rising fast and their internal supply flow is shrinking.

CS: Right. But you said something else interesting there that I'd like to ask you about. I would think that it would be in the interest of OPEC to lie about this, insofar as the moment that even the skeptics in the United States realize a) the reality of peak oil and b) that we are right there—even the oil people are going to run for renewables.

MS: Well, see, the problem is that we can't run for renewables because there really aren't very many that we know of that will actually change the bar. But what is interesting is if the world believed that OPEC had 4 to 6 million barrels (per day) of excess capacity they are just shutting down and we all of a sudden had a tremendous price rise and shortages, we'd bomb OPEC to get their oil.

CS: Wow. That's quite a statement.

MS: You could almost make book on it. Because you could say if those, and I'm being particularly pejorative because it's the way we would think, if those greedy Arabs, and I still vividly remember John McCain saying "The A-Rabs have all this oil and we should not let them use it." And we would have schizophrenia about the Arabs and sooner or later somebody would say, "We need to bomb them and get the oil because we are ruining our economies."

CS: Well I personally believe that, but I must say that I'm surprised to hear it from you. The concept that we go to war for oil is something that I think a lot of Americans believe in our hearts, but I don't know that there is any proof of that—and it is certainly vehemently denied by whomever we have in power.

MS: Michael Klare, a very good writer and documentarian from Amherst, has a documentary movie out called "Blood and Oil" and it showed how from World War II on, every single administration has acknowledged that our wars are always fought over the oil.

CS: Okay. Well that's good to know. I'm glad we got that out on the table.

MS: See, Mexico, for instance, is one of the few countries that decided they needed to be transparent in the field-by-field flow rate and be very clear that Cantarell was in decline so that people understand and sympathize with them. Then they could finally say, "Within a year or two we can no longer export oil to the United States," as opposed to thinking that some greedy Mexican is trying to get the price up.

CS: I understand. Well, speaking of manipulating the price, can you speak to that for a second? Many people attribute the 2008 skyrocketing gasoline prices in the United States to manipulation.

MS: Over the course of 10 years—from the fall of 1998 when oil prices were at 10 to the early summer of 2008 when prices were over 140.

CS: Right. In dollars per barrel.

MS: That's a long, long, period of time to have prices rise that high. And what happened is that demand far exceeded supply. I mean, throughout the price rise it was blamed on speculators and manipulation and so forth. You don't do that for a decade.

The stranger thing was that in the fall of 2008, the price had sort of come off as we started worrying about our weakening economy. And on September 22nd it was $122 a barrel, and on December 22nd it was $32 a barrel. That was the oddity. We should say, "How did that ever happen?" And then, surprisingly enough, it took three months just hanging around the $30 level before we had the two big rises. It went back to $60 in four months and then it stayed there for two months and then it went up to $80 in the next four months. So in 2009 we had the highest single rise in the price of oil in a single year in the last 40 years.

CS: Right. Both percentage-wise and...

MS. ...and in total dollars. And now we have the worst cold weather we've had in the Northern Hemisphere in decades, and my guess is that oil prices pretty soon will be back over $100.

CS: Okay. Let's go back to the subject of peak oil. If you don't mind just documenting the proof points of this thing for the skeptic?

MS: Well, the proof points begin with a number of very important key oil producers. Nobody's data is perfectly accurate because of the way we measure oil flows, but take the North Sea's for instance. This was the last great significant new frontier; between the two big producers in that area—Norway and the UK—in 1999 the North Sea produced 6.1 million barrels a day and today it's down to about 2.5. Mexico's Canterell field peaked in 2005 at 2.2 million barrels a day and today it's 500,000 barrels a day. The North Slope peaked at 2 million barrels a day in 1989 and today they struggle with about 600,000 barrels a day. So you can go through enough of these individual countries and say, "How would we ever basically just replace the Cantarell and the North Sea?" The truth is that we can't.

CS: Do you mind addressing the position of people who say, "Well there is so much oil in shale and tar sands and so forth?"

MS: Right. Unconventionals. The problem with the unconventionals can be summed up in a single word: they're unconventional. And they cost an enormous amount to turn them into flow-rates.

They also use a remarkably high amount of water—and often times other energy. The oil sands of Canada use just a phenomenal amount of potable water and natural gas to actually steam it out of the sands. In California, something like two thirds of their oil supply comes from Kern County's heavy oil, and while the San Joaquin Valley is one of the key food supply sources of America, they're having a very serious drought.

And yet what are they doing? They're basically taking potable water and natural gas to create steam to do steam soaks to get heavy oil out of the ground that then needs to be refined about four times more than a typical light oil would do before it's usable. So we ought to ban that stuff because basically it's an energy destroyer.

And I think as we move into the future, we are going to end up becoming far more alarmed about our water scarcity then we are our oil scarcity. And people are going to start to get more educated on how much water we consume to create usable energy.

CS: That's exactly what I was going to ask next. That is, this seems to take its place in an entire constellation of shortages and scarcities. Please go ahead.

MS: The worst shortage is always water, because without water we die. And I wasn't aware of some of the water statistics until the last six months when I finally waded in and started reading two or three of the best books out on the subject. Planet Water is a really first rate book, and Blue Covenant, which I finished over the holidays, is another first rate book.

And I would definitely recommend Cadillac Desert, which was written in 1985 about water scarcity in the West. 25 years have gone by, and we are headed into some water scarcity issues that are really scary, but look at the incredible amount of water that we used to basically create usable oil. One of the statistics that surprised me the most, in fact ironically just as you called I was asking our refinery analyst to check and see whether this number is in the ballpark of being accurate, is that a refinery uses 21 million gallons of water an hour.

CS: That's amazing.

MS: It is amazing. So you think of the amount of water that's used to extract oil. A lot of the major oil fields are doing water injection to basically create artificial reservoir pressure to force the oil out of the ground, and then when you get to refine it, we just chew up water to get oil. And the water basically has no cost to it. Well, sooner or later, you're going to insist that the oil companies pay for the water. And it will probably triple the price of oil.

CS: Wow. How interesting. What would you say are the consequences of peak oil?

MS: The biggest consequence is that we have developed a society that is addicted to oil. And the poster child of this society happens to be China, who we're just getting started on their relentless drive to become like us. And then right behind them is India. Well if India and China ended up with no population growth, which is impossible, and ended up finally climbing up the ladder and consuming as much oil as Mexico does today on a per capita basis, which is about a fifth of what the United States does, it would take another 45 million barrels a day of oil. Which is obviously impossible.

CS: Right. And what's the production right now in terms of millions of barrels per day?

MS: Crude oil is 72 million barrels per day. Total petroleum is about 85. So we are squeezing out the rest through a bunch of miracles. The 72 is going to drop to 60 within the next 5 to 10 years. While we at the very minimum have a world that is expecting to use 105 to 120 million barrels a day two decades from now.

CS: I understand that we've become addicted to oil in a very real sense of the word, but I guess my question is what are the consequences of that addiction—in terms of international relations, human suffering—that kind of thing.

MS: The consequences are pretty simple. We don't have any way to basically regulate the use, because the users of oil have no idea of when we're almost empty. It's just like—have you ever run out of gas?

CS: Yes.

MS: Don't you feel stupid when you run out of gas?

CS: Yes, in fact I do.

MS: I've never asked anyone that who has never run out of gas, and I've never had someone say "No, no, I knew I was going to run out of gas". You just forget to look at your gas gauge. We have no national—let alone global—gas gauge. And at some point, our motor stops and we're where we were right after Hurricane Ike. We had about 2 1/2 weeks of service station outages spreading from east of Houston all the way down to the middle of Florida and up to Baltimore. If we had that happen again, and we hadn't had the financial crisis going on, it would have been the lead news event every night and then the motorists would have topped off their tanks. That's just the way we clear the shelves of water when hurricanes approach. And within 30 hours, we would have drained the usable gasoline supplies around the service stations of America and we would end up with social chaos. And within a week we'd be out of food.

CS: So here's a question for you. I'd like to end by asking what we should do. In particular, can we ameliorate this with renewable energy? You seem to be saying that we are a long way from renewable energy.

MS: Well, most of the forms of renewable energy have one of two problems. They either take decades to scale to where they'd make a difference in size or we basically don't have the technology to do it. There is a laundry list of renewables that people are talking about—and certainly the Obama administration has been on the warpath to create a lot of these targets—but none of the numbers work.

CS: You know, there are people who would disagree with you.

MS: Yeah, I know, I talked to a lot of them. But they don't have their homework.

CS: I'm sure you've seen stuff on solar thermal—concentrating solar power—the Ausras and Abengoas and so forth of the world. If were betting on a technology—I don't think it'll be here two years hence but I would say 15 or 20 years, a solar thermal farm the shape of a square 100 miles on a side would give us more than enough power for the entire continent of North America.

MS: Yeah, if we had superconductivity. See, the problem is I don't think realistically we'll have electric vehicles that will replace even 10% of the vehicles in the United States within the next two or three decades.

CS: Really?

MS: Well look at the experience of Toyota. A year ago last summer, Toyota finally sold its millionth car, a Prius. And it took 10 years. It took the previous 10 years to design it. We have 280 million vehicles in the United States. So 1 million electric cars, let alone 10 million, just don't even end up as a spit in the ocean.

CS: I do understand that. But I'm a little miffed with the auto OEMs for not moving on this thing. Toyota's a wonderful example of a company that was already considered to be green with its Prius and could have put a plug hybrid on the road eight years ago if they wanted to. But there just simply was nothing in it for Toyota.

MS: See, in my opinion, the turbo diesel was about four times a better car than the hybrids. And diesel is easier for a refinery to make. I drive a 320D Mercedes that's about three years old now, and the new model's apparently almost 30% more energy efficient. Driving around Houston in stop-and-go traffic I get about 25 to 30 miles per gallon. On the open road I get 40 to 45.

I think we can end up where light rail works through electricity. That's the closest thing we'll get to using electricity to transport people.

CS: Right, and what about the Nissan LEAF? Maybe I'm just drinking the EV Kool-Aid here, but it seems to me that we're just a couple of years away from production scale small EV's. Basically every OEM: BMW has the Mini E, Mitsubishi has the i-Miev—everybody's trying to do this.

MS: But the problem is, if it worked better than everyone thinks, realistically could we really think that we would have 40 million on the highways within two decades?

CS: Sure.

MS: Well, if we did how does that compare to the 900 million vehicles we have on the road today?

CS: Well yeah, right. I'm talking about the United States—but let me just cut to the chase and ask: What is the answer?

MS: In my opinion, the answer is something that I'm heavily involved in now in Maine which is called the Ocean Energy Institute. Our big project that we have under way is to create in the Gulf of Maine the experimental proof that you can turn offshore wind through electricity into liquid ammonia and have that replace motor gasoline and diesel fuel and jet fuel.

That's something that actually works in internal combustion engines. So once we've proved this, you can see within five years maybe 2000 of these turbines around the coast of China and Brazil and West Africa and the Persian Gulf, so the future is liquid ammonia.

CS: OK, well that's very interesting. Offshore wind—a lot of people are talking about this.

MS: The University of Maine has created an advanced composite that's the strongest, lightest material ever made. And it has almost no energy content.

CS: Do you know what it's called?

MS: "Advanced composite." That's what they call it. It's a trademarked secret, but it's basically resins and sand and sawdust and it's just a remarkable thing, and they've made bridge spans crossing the Penobscot River, so they stress-test these things. In three months they put 80 years of truck traffic simulation on it and I've seen two people hold up the bridge span. So this stuff actually works and this will allow in the Gulf of Maine in deep enough water to use a certain design to create huge wind turbines every 5 miles apart.

With all these alternative energies I hear people talk about, liquid ammonia is the one everybody has missed. And I will have to say, this is something that I did my research on and figured out it was going to work.

CS: Well great. So then do you consider us safe? Do you think that this is a bankable process?

MS: No, I think that we're going to have to go on an effort with the intensity of the Manhattan Project, like our work after Pearl Harbor. We have about five years to basically figure this out and prove to the world that we'll be cranking out the advanced composite wind turbines like we did liberty ships. Go to the website: Oceanenergy.org

CS: But where is all this effort going to come from? Certainly not traditional energy companies, I would think. I don't want to be crass about this thing, but it seems to me that if you are Chevron saying, "Imagine an oil company being part of the solution" you're talking to people who simply don't believe you. They are not part of the solution; they're trying to milk to pump the last ounce of crude out of the ground.

MS: That's very accurate when it comes to Shell, BP and Exxon. Chevron is almost bold enough to break out and say that they actually acknowledge peak oil. I had never met Jim Moulder—the chairman and CEO of ConocoPhillips until the "Oil and Money" conference in London this fall. He was so friendly; he said "You know I'm just embarrassed. I kept dying to just pick up the phone some time and call you and get acquainted because I love what you write."

So I said, "Well, when I'm back in Houston, let's get together." We spent 2 1/2 hours talking. He said, "I just don't think there's a big rosy future for the oil business and I kept hearing people telling me that shale gas was the answer and I was so impressed to hear you speak up in London and say 'No, shale gas is a bunch of crap.' So first of all I'd like you to tell me why you think that." Thirty minutes later he said, "Holy cow, I'm glad I didn't get into that." But finally he asked, "Is there anything you see on the horizon that is exciting? This is pretty depressing." I said "Oh yeah, offshore energy, offshore wind, liquid ammonia." And within half an hour he said, "This is actually the most exciting conversation I've had in my business career."

CS: How interesting. This would have been a much less important book, had I not had this conversation.

MS: Well, I'm glad you're doing it. We need education so badly on this issue. What's most interesting to me is the convergence of water scarcity and oil scarcity; it's a real killer.

CS: Okay. Matt, I can't thank you enough. This has been a fantastic experience for me and I'm certainly looking forward to meeting you.

MS: Excellent.

For more information on this contributor, please visit:
http://2greenenergy.com/renewable-energy-facts-fantasies/.

ENERGY AND NATIONAL SECURITY

James Woolsey has received a total of four presidential appointments—under both Republican and Democratic administrations. Most notably, he was Director of the United States Central Intelligence Agency from February 5, 1993 until January 10, 1995.

Mr. Woolsey is among the most vocal and, in my view, most credible proponents of renewables, making arguments touching on national security, global climate change, and economics. He is featured in Thomas Friedman's Discovery Channel documentary Addicted to Oil, *and in the 2006 documentary film* Who Killed the Electric Car? *which addresses solutions to oil dependency through the development of electric transportation. I was elated when his assistant responded positively to my request for an interview.*

Craig Shields: From what I've learned from your talks and writings, you see numerous imperatives to get off fossil fuels, but when we look back on the last several decades, we see a lot of people sitting on their hands. Why do you think that is?

James Woolsey: Well, there're two aspects of energy and security as far as I'm concerned. One is the security of the grid. For us that's not a supply problem. We don't have the European-type problem of having some guys who might act up like Putin controlling the fuel for our electricity; we pretty much make our own electricity. But the aged nature of our transmission grid produces several big national security problems, including vulnerability of the transformers and other things to terrorist attack, hacking in the SCADA systems, and EMP (electromagnetic pulse). With countries like North Korea and Iran probably getting nuclear weapons and being able to just launch something from 200 miles off the coast—200 miles up—an EMP could take out a huge share of the grid.

So, you've got those grid vulnerability problems for the electricity part of our national energy set-up, and then you have vulnerability with respect to oil. You could say that the supply network is potentially vulnerable; if you take out the right buildings in Louisiana, you could really screw up control of the oil pipelines. So generally speaking, you have concern about terrorist attacks. But oil has a whole range of national security problems associated with it that don't really exist for electricity. So, I think there're two big segments of the energy structure, electricity and transportation, that have slightly overlapping but somewhat different types of national security vulnerabilities.

CS: Yes. As I like to say, I know there are people who don't believe in global climate change. But are there people who don't believe in terrorism?

JW: Sure, oh my, you've got a whole bunch of things. You've got, on the pollution front, you may have already seen Boyden Gray's piece in the Texas Review of Law and Politics about three years ago on the aromatics. Boyden puts the cost in damage to peoples' health and medical costs total, at approximately $250 billion a year from the aromatics.

CS: Yes, the externalities, quantifying the cost of lung disease and so forth. There was an op-ed in the New York Times just the other day that I thought did a great job with that.

JW: Right. So you've got damage like this that's unique to oil, which is not really normally thought of as a national security issue, but you've got both the terrorism and enhancing of the bad guys. What Tom Friedman calls "Fill 'er up with dictators," and all the issues associated with that.

CS: Let's talk more about empowering the bad guys—especially insofar as we have such a moving target in Al'Qaeda. By the way, you're speaking with a loyal tax-paying American who really wants to be on the right side of this. But I also represent the people who don't want to be at war unless there's an incredibly compelling case to do so.

JW: I don't think the main issue is that we have a larger military because we have to protect oil. I mean, we weren't in Bosnia about oil. We weren't helping in Kosovo about oil. We weren't trying to feed the Somalis about oil. Yes, it's true that we might have ignored Saddam's conquest of Kuwait. You recall some wag said at the time if Saddam—instead of coming within 100 miles of controlling over half the world's proven reserves of oil, he'd come within 100 miles of controlling over half the world's reserves of broccoli, we would have stayed home.

CS: Right, I do remember that.

JW: So there's a strategic element there, but I don't think that one can go through the armed forces and say that we can do with one or two fewer battle groups if we just didn't have to worry about oil. I'm an old Scoop Jackson democrat and tend to never see a submarine or an aircraft carrier and not want an extra one. But I don't think that's the issue. I think it's several things mixed together. First of all, oil, like gold before it, has the effect that Paul Collier at Oxford, and Tom Friedman cite sometimes called the "oil curse." Generally it's just that an autocratic state, when it depends for a huge share of its income on a commodity that has a lot of economic rent attached to it, that rent accrues to the central power of the state essentially. So you tend not to have representative institutions like legislatures and you tend to have a much more difficult time getting out of an autocratic structure than with a broad-based economy.

If you look at evolution, the examples I usually use are Taiwan and South Korea. They were tough dictatorships, but as they prospered and built up a middle class—and this happened to them a lot faster than it happened in Europe in the medieval and early modern times—it was a similar phenomenon. The middle class builds up, it's diversified, it starts wanting economic liberties and that transmogrifies after a while into political liberties and it tends to gravitate toward freer institutions. That tends not to happen when you've got a lot of economic rent associated with a commodity that you're heavily dependent on. Read Larry Diamond's book if you haven't already. If you look at the 22 countries that count on two-thirds or more of their national income from oil—it's fair to say all 22 of those countries are autocratic kingdoms or dictatorships.

And I haven't compared that list with Freedom House's list of the forty basically—those that Freedom House calls "Not Free." There are about 120 democracies in the world, I mean not perfect, but nonetheless regular elections and another 20 countries like Bahrain that are reasonably well and decently governed, even though not democratically so. And then you've got 40 really bad guys. And I'm pretty sure that list of 22 in Larry Diamond's book is virtually all from the list of 40 bad guys—or "Not Free," in Freedom House's terms.

So you've got that effect, which is, like anything in this area, not a clear bright line, but the countries that export a good deal of oil like Canada and Norway that are clear democracies are not in this category of two-thirds of their national income depending on oil.

CS: Right, I'm with you.

JW: So it's really a pretty decisive set of statistics, I think, and then if you look at other numbers, set out in places like Mort Halprin's book The Democracy Advantage, it's pretty clear that basically democracies don't fight each other. They occasionally get really pissed off, but they mainly choose up sides and argue about trade sanctions and stuff. It's not impossible but it's really hard, even going back into the 19th century, but certainly since 1945, finding democracies fighting each other. They just don't.

So you've got oil locking some states that depend so heavily on it into autocracy and dictatorship and worse. And those are the folks who also fund the terrorists, who invade neighboring countries, etc. So there's a large national security point here—but that's not the way people often talk about it. We often put it in purely American terms, but I think it's really bigger than that. It's a big problem for us because we tend to end up being the world's policeman and so forth, but it's a problem for everybody.

Now, here's a second point on that. Read Alex Alexiev's cover story in a relatively recent issue of the National Review, and Lawrence Wright's *The Looming Tower*. With a little over 1% of the world's Muslims, the Saudis control about 90% of the world's Islamic institutions. Now, given the character of Wahhabi Islam, as contrasted to something much more open and generous—even set aside the radical Shiites in Tehran, and just look at Sunni and Wahhabi. An example I've used on a number of occasions is if you took Ferdinand and Isabella and Torquemada and moved them up into the 21st century, put 25% of the world's oil under Spain, and gave Torquemada $6 billion a year in order to spread the Spanish Inquisition.

I've also drawn a parallel between the Stalinists and the Trotsky-ites. They both wanted to knock off the bourgeoisie and establish a dictatorship of the proletariat; they just disagreed over who should be in charge and whether you ought to be able to go off like Trotsky wanted to and start revolutions wherever you wanted, which is kind of like Al'Qaeda, or whether you stayed with the disciplined structure and obeyed one leader, which is the Wahhabis and the Stalinists.

CS: This is fascinating. Let me ask you this. Looking at the US and the imperatives both domestic and foreign with respect to oil, we look at the last few decades of our history and you might say, Well, during the embargoes of the seventies, during the Carter administration that ended in 1980, it looked like we were getting our act together with respect to renewable energy. Starting right at the end of the Carter administration, we turned around and ran 180 degrees the other way. Is this true?

JW: It really was 1985. The Saudis dropped the price of oil down to close to $5 a barrel and all of these ideas, whether they were good ones like bio-fuels or bad ones like Synfuels Corporation, pretty much all got bankrupted. I think the Saudis were probably mainly going after the Soviets and it did have a devastating effect on the economy of the Soviet Union; I think it had something to do with the system collapsing four years later. But they also basically bankrupted all the efforts to come up with alternatives.

And that lasted into the '90s. A little bit of stuff got started in the '90s, but in the late '90s they took it down again. This was probably mainly the Asian recession, but it happened. They took it down to close to $10 a barrel and a lot of the work that was getting going on cellulosic feed stocks for bio-fuels and stuff like that pretty much stopped. And then, shortly after it came in, the (George W.) Bush administration decided it was going to push hydrogen—which was, I think, a particularly dumb decision.

People who've wanted to come up with replacements for oil have either gotten trashed once a decade or so by the Saudis, just exactly the way John D. Rockefeller used to trash his competitors, or they've gotten off on crazy tangents. A few actually believed that hydrogen would be a great transition, though some of that was just Detroit figuring out a way to spend a few million dollars and get a lot of publicity and not get bothered by having to do things like really improving fuel economy or putting out flexible fuel vehicles.

So, it's been a mixture of things, but in a way the problem comes down to the fact that petroleum absolutely dominates transportation about 95%-96% around the world, I think, and OPEC dominates to the tune of three-quarters or more of the world's proven oil reserves. So you can get an argument about peak oil and whether they could still do it or not, but the Saudis, at least in the past, have pulled a John D. Rockefeller a couple of times, inadvertently or purposefully, and that's been a good chunk of the reason that people have gotten discouraged and turned away from it. There is a certain character to the American people being like Charlie Brown trying to kick the football every fall as Lucy pulls it away from him.

CS: It's impossible to see into the future, but it is possible—especially from your perspective—to analyze the past. Now, when you say, "There were probably a few people who sincerely believed in the hydrogen economy and that this was the technology that would take us away from oil," you're implying that there were people who weren't sincere about this at all.

JW: Yeah, I think there was a real bait and switch aspect to it.

CS: Do you mind expanding on that?

JW: If whenever anybody says, "Don't you remember the 1970's and don't you remember the oil cutoffs and don't we have a problem with foreign control of oil," and you say "Yes, but the wonderful solution is the hydrogen economy and hydrogen fuel cells for cars and look at our car here—but don't ask how much it costs. Yeah, it was 1 - 2 million dollars, but price will come down and we'll have beautiful glossy prints in magazines and we'll say to the world that we are dealing with the problem of oil dependence by going to go to clean hydrogen." I think that was the pitch—and some people believed it, probably, but I think that for quite a few it was just a way to keep looking like you're doing something and it's a lot cheaper to have $2 million cars and one or two hydrogen stations that you can have pictures taken at, rather than to figure out exactly how in the hell you're going to have a nation-wide infrastructure of hydrogen fueling stations for anything less than many hundreds of billions of dollars.

CS: Right. And I'm not trying to be provocative here, but I do want to ask you this. There are some who would say that we elect leaders to act in our interests, not in theirs—and if you have, factually, leaders who were oil-men and who knowingly threw out a red herring in the form of hydrogen, thus profiting themselves while subjecting the rest of us to great harm and danger—not only heath-wise, but also security-wise—if it's not treason, it's pretty close.

JW: Well, I grew up in Tulsa, which fancied itself in the 50's, and even into the 60's or so, the oil capital of the world. And it was probably true, back in the 20's. The oil business is still there, and pipeline companies and so forth are still a huge part of Oklahoma's economy—and Tulsa's in particular. So I grew up in the middle of this. And the oil business always regarded itself as not understood and not appreciated, but they were the guys who went out there and really drilled and got the stuff that was necessary to run the economy. And they saw themselves, and some of them still do, as being unappreciated—doing the hard and grubby and dirty and important work of fueling the world's transportation and a lot else besides. And that mindset, quite apart from any cynicism, is present in a lot of people who work in the offices.

Second there's a lack of willingness to regard the market as being as heavily controlled as it is. I heard a guy from AEI (conservative think-tank American Enterprise Institute) the other day at a conference had talked about open standard flexible fuel vehicles, so you could run them on ethanol, methanol, any mixture, including gasoline, etc. And he was very much of the view that the government shouldn't be interfering with things by requiring flexible fuel vehicles. Well, in a manufacturing process, the last I heard, they're about $35. There's a hundred dollar oxygen sensor in there that you need; if you really look at what it would cost in the manufacturing process to make these FFVs, it's a different kind of programming in the software and a different kind of plastic in the fuel line. (Venture capitalist) Vinod Khosla has made a special study of it, and it's about thirty-five bucks. So what this guy from the think-tank was saying, was it's too much of an interference with the market for the government even to require a $35 part. I mean it's a tiny fraction of what seat-belts cost for example, or airbags. If it was important, the market would do it. There are a lot of people who think that way.

Now, if you look, it's kind of interesting in the first G.W. Bush administration, they were very much into hydrogen and all this stuff, but Bush himself came around on oil. I don't remember if it was early '06 or early '07 where he went through the business of "addicted to oil"—and in some of the appointments to the DoE, they got rid of a guy who really was not interested in renewables who was heading up the renewables office. They put in a very good appointee to work on renewables...and they kind of shifted gears a bit in the last couple of years. I think it was just largely a matter of overcoming some of these mindsets about the market and the quasi-heroic role of the oil business. It's a complicated thing.

Craig, you didn't ask, but I don't think it had anything to do with Bush's or Cheney's personal interests or anything like that. I don't think that's what happened. I think they were off on these historic tangents. Bush finally started getting interested in electricity, electrifying transportation, but until relatively recently people weren't talking about electricity as a way to get off oil for transportation. They were talking largely about ethanol. And although I think the problems with ethanol are not nearly as severe as the Grocery Manufacturers' Association and the oil companies have said, there are some difficulties. It can't be used in pipelines, needs a separate pump, etc. So it wasn't like finding out that your stomach doesn't agree with milk products and just going to the store and switching to soy; you couldn't do that. You needed to go through a number of steps, and I think those several things kept them focused on oil.

I guess the final thing is that you really do help on the balance of payments; it's the only thing you help with, by going to more domestic production. If you are borrowing a billion dollars a day—which is about where we are with seventy dollar a barrel oil—if you're borrowing a billion dollars a day for imports, every 365th of your imports that you can produce domestically instead of importing, you save a billion bucks for the balance of trade. So you do that. I mean "drill, baby, drill" has something to it. But it doesn't solve the fundamental problem of oil dependence, which is all this business about the foreign dictatorships and the Wahhabis and all that I was talking about earlier.

CS: Right, not to mention global climate change.

You mentioned this thing about the $35 part. I could understand that a rabid free market economist—a true Libertarian—might say let the market decide. But most people, I think, would probably say, "Look, as a civilization, there're almost seven billion of us. If we really wait for true market economics to take hold here, we'll all be dead."

JW: Oh, I completely agree. I mean I think that's far and away the better argument. I was just trying to put into context the various things that could reasonably be in the minds of somebody who didn't want to have the government take steps to get off oil. I think the government ought to get in there and basically bust the trusts—a Teddy Roosevelt approach. I think that this is what the government ought to be doing. But the two times that the government has made a choice of where to go, it was the Synfuels Corporation and hydrogen—and neither one of those was very wise, to put it mildly.

I mean I think Carter thought he was doing a good thing getting the Synfuels Corporation started. Now it of course put a huge amount of carbon into the air and it was extremely expensive, but they weren't thinking about carbon back in the 70's—not many people were anyway. And the expense was something they thought would be bearable because oil was always going to be up there at many, many tens of dollars a barrel.

CS: Great. Well this has been fantastic.

JW: Good talking to you.

For more information on this contributor, please visit:
http://2greenenergy.com/renewable-energy-facts-fantasies/.

GLOBAL CLIMATE CHANGE

Professor V. Ramanathan of Scripps Institution of Oceanography is the man generally credited with the discovery of the phenomenon of global warming, correctly predicting in the early 1970s that there would be a measurable increase in the temperature of the Earth's atmosphere by 1980. I was happy to make the 200-mile trek down to La Jolla to speak with "Ram" (as he likes to be called) in his laboratory.

Craig Shields: I thank you so much for having me here. Maybe we could start with the early days. Could you take us back there and tell us what it was like?

Dr. Ramanathan: I was fresh out of graduate school. Earlier, I had studied the atmospheres of Mars and Venus, and for my Ph.D. I wanted to switch to looking at the more pertinent environment—something more relevant. At that time my main interest was carbon dioxide, which comes from fossil fuel combustion, and how much it was contributing to so-called "global warming."

There was a paper that was looking at the effect of chlorofluorocarbons that pointed out that CFCs would go into the atmosphere and destroy the ozone layer—this paper ultimately won the Nobel Prize for Chemistry. And I became interested to see if they had a greenhouse effect. And to my surprise I found that one molecule of CFC had the same global warming effect as adding 10,000 molecules of carbon dioxide. My paper was published in 1975, and it was met with skepticism and disbelief; it took scientists five or ten years to re-do my work and convince themselves that, yes, this was a major issue.

But in any case the discovery of the greenhouse effect of chlorofluorocarbons had opened Pandora's box. Soon, other scientists and I discovered that a whole host of gases we are letting loose in the atmosphere are probably more urgent than what we thought about CO_2 increase. So I got curious, and in 1980 I published a paper along with another famous meteorologist Robert Patton, in which we concluded that if the hypothesis or theory of global warming was correct, we should see this warming by the year 2000. And of course a team of over a thousand scientists in 2001 agreed with the idea of global warming from greenhouse gases, and that prediction was confirmed.

That was good for the science, but not good for the planet because it continues to get worse. From the time I first published my paper and we discovered this greenhouse warming in the atmosphere, we had released billions and billions of tons of these greenhouse gases.

CS: I understand that the oil companies are still spending money trying to discredit this. Do you ever feel personally attacked?

VR: No, it was not an issue of a personal attack. When you publish an idea, people have to be able to repeat what you did and have to be able to see the effects in the atmosphere. I feel more sad that the planet is heating, and that the warming you've seen is much larger than even I had predicted. In fact this early warning is being ignored; we are still just dumping pollutants, with enormous disastrous consequences 20 or 30 years from now. You are not talking about something that is just going to happen in 200 years; you are talking about something that will happen in 20 or 30 years. So this is the time to do something, not to be bickering. That is what makes me sad. We are losing valuable time to fix the problem.

CS: We are just a week or so away from Copenhagen. What should a rational mind think going into Copenhagen? Do we have reason to be optimistic about people coming together and stopping the bickering— or do you think that this is going to be an exercise in verbosity for its own sake?

VR: The good thing is that almost all the leaders—including those from the developed and developing nations—agree that we have to do something and we have to do it now. The issue with the leaders is that everyone is waiting for somebody else to take the lead. The developing nations tell us, "You caused this, you fix it," and the developed nations are telling developing nations, "Unless you join us we are not going to do it." So in a sense that is the issue, but at least we are past the first major hurdle that everyone accepts it is a major problem. And if we do not fix it then it will have disastrous consequences. So we are past that hump at least.

CS: Yes, I think most of us are. I was telling a friend the other day that if you had asked a hundred people a year ago what degree of credibility the concept of global climate change has in their mind, you would have batted about 99 out of 100. I am not so sure that is true today. I think there are a significant number of people who actually doubt the hypothesis of global warming.

VR: That is true. I think it led to a crescendo a year or two ago, and now we seem to be siding a little bit more with the skeptics. It is a problem and I cannot deny that. But why is it happening? I think it is economic catastrophes, people can only take so much bad news. It's hard to know how much the propaganda of the oil industry is hurting this, but I see some commercial advertisements on TV by the oil industry promoting alternate energy. But I think the key thing is that I have not seen one leader get up and say, "No, we don't have to worry about it." At least we are past that stage. But we need enormous education for the public; that has to happen.

CS: I read a couple of the articles on your website and I am interested in this phenomenon of "runaway events." So for instance, the likelihood that we are going to lose the permafrost, which is burying the methane. Could you comment on runaway events like that and others and the probabilities that are assigned to those? In particular, I've read papers that are extremely precise about the probability of a certain specific phenomenon. I am wondering how can someone assign, with any real certainty, a probability associated with a one-time event in natural and geologic history.

VR: Let us take the situation where we just continue dumping these gases at the same rate we have been doing. My own work suggests in less than 30 to 40 years we will cross the threshold of two degrees (Celsius) warming. The planet has already warmed about three-quarters of a degree. So if it crosses this two-degree threshold, or in Fahrenheit it is working on close to a 3.5-degree threshold, I expect to see major iconic changes in the planet. For example, the first to go is the arctic summer sea ice; the Earth would look bluish, not white, when you look from space into the polar region. Now that has got a huge impact on the ecosystem, on temperatures, on global warming, etc. The second major disastrous thing—you can think of climate "tipping points" in what is happening—is the melting of the Himalayan glaciers—those glaciers provide the head water for all the major river systems in Asia. Three billion people depend on that. And the third is the permafrost, as you mentioned, evaporating, melting away and exposing the methane. These all one by one will create the climate tipping point, with major disastrous consequences. At this stage, our understanding of this is not such that we can say definitively what is going to happen. Yet, we can give a probability of these things happening and there is at least a 50% probability that any and all of this can happen if we pass that two-degree threshold.

CS: Is this specifically tied to 385 parts per million CO_2? In your estimation, is there a specific level of CO_2 that would trigger one of these events—or do you think this fixation on CO_2 levels is too narrow?

VR: The first thing I need to point out is that CO_2 causes about 50% of the trouble. The other 50% comes from other gases like methane, we talked about fluorocarbons, nitrous oxide which comes from fertilizer, and black carbon. So I do not see it as an issue of just limiting CO_2; we need to look at the whole picture. And the carbon dioxide concentration is already 380 parts per million. So we are already there and so an answer to your question is 385 narrow, my first answer is that 380 does not tell us what else is going on. We have to reduce CO_2, but we also have to reduce the cause of pollution in other gases. And even if Copenhagen succeeds in cutting down the CO_2 emissions by 50%, the carbon dioxide concentration will exceed 400 parts per million. So we are already going to be past all these thresholds.

So the question then is what can we do. There are so many short-lived greenhouse gases; some of us are promoting so-called "fast-track" action. Just to give an example, one of the greenhouse gases is ozone. Its lifetime is only a month, so if we cut down the emission of the gases which lead to ozone, they are gone within a few months. So we can do a lot of things, and not be blind-sided by just focusing on carbon dioxide. We have to cut down on CO_2; do not get me wrong on that—but that is not going to save us from exceeding this two-degree threshold. We have to cut other greenhouse gases too.

CS: I am interested in masking and enhancing effects—i.e., in phenomena that serve to speed the warming process up or slow it down. So we have aerosol, soot, and so forth. We have components in the atmosphere that your UAV (unmanned aerial vehicle) project is attempting to quantify. What would you say is the net of all these masking effects?

VR: It is best to think of the greenhouse gases as a blanket. Just like as a blanket traps our body heat and keeps us warm, this blanket of greenhouse gases traps the heat coming from the sun. Some of these particles in the atmosphere function to warm or to cool. We really do not know the true nature of the beast in terms of a greenhouse blanket because of the amounts.

CS: But your guess is what? If you were a betting man and you were saying "I bet that the totality of the mirror effect and the enhancing effect, the black body effect, call it what you will …. is what?"

VR: My own estimates show these mirrors have masked about 50% of the warming effect by the blanket. So when we clean up the air and get rid of the minerals the warming we will see will be at least two degrees already.

CS: OK. Obviously renewable energy sounds like a good thing, especially based on this conversation. But I would think burning,—even biofuels, would be an inferior solution to solar or wind, or geothermal, etc. I would think if you had the choice between solar thermal and biofuels you would chose the former everyday because the latter burns hydrocarbons. Talk to me, if you would, about renewable energy and where would you like to see us go. Maybe take this on a continent-by-continent basis—there might be something different we need to do in North America versus say Europe or Asia.

VR: Solar to me is a logical first choice and I can give you a personal example. I switched my home to solar completely. And it will pay back in eight years, and then the rest of the fuel that it produces is free. So I do not see the argument that solar is costly. It is there now, the problem is solar is not going to work in Alaska. So I would say certain latitude regions, between 40 degrees north and 40 degrees south, solar is there. The main problem with that is the capital cost; for example for a home like mine, it was $14,000—and that is a lot of money for some folks.

And solar is stationary; it is not going to solve the transportation issue. We need a whole array of solutions. I would say for power generation I would go for solar all the way. I have understood the argument that it is expensive. An issue with solar is storage—how do you store the energy. And as far as biofuel is concerned, things like corn ethanol is a disaster, simply because you have to grow corn first, which means you need water, you are coupling energy with water, you do not want to do that. Then you have to use fertilizers and that produces other greenhouse gases.

The most promising thing with biofuel is the marine algae, but that technology is not there. And sometime in the future, whether that future is two years from now or 20 years from now, we will see. There are some solutions we could start, energy efficiency is clearly a logical choice. I think we have gone past the time when we could say, "I will choose one". That luxury is gone. We are losing time.

CS: I can hear the passion in your voice. And I understand it completely. Here you have 6.8 billion people whose lives are at stake. Tell us a little bit about the emotion with which you approach this subject.

VR: I spent the last 35 years working on the science. And my work was producing one bad piece of news after the other. And a few years ago I decided I had to become part of the solution, not part of the problem. So first off that is reducing my own carbon footprint. I know it is difficult; it is not easy because we are all addicted to fossil fuel. On the solution side, I discovered just starting to work on this, fortunately it is not to late. We can and must do a lot of things. Just to give you one simple idea I am pursuing, roughly three billion people in the world have no access to fossil fuels, so they are basically burning wood, and firewood which releases a lot of carbon dioxide greenhouse gases and other forms of black carbon. So it turns out that giving them access to cleaner fuel and cleaner ways of cooking would save a tremendous amount of these climate change pollutants. So the solution is there; we have stuff that does not pollute. There are a number of these solutions waiting out there, what I call low-hanging fruit, and we have got to go for that.

CS: Excellent. We have talked about this a little bit before. Chevron famously goes on television and says, "Imagine an oil company being part of the solution." Well I am sorry I cannot. I flat-out do not believe that. What do you think, if you do not mind making kind of a quasi-political statement?

VR: Interestingly enough, I've written papers on global warming which, of course, no one knew about until 20 - 30 years ago. I know the attitude the oil industry took was extremely skeptical and negative. Now, take the example of Chevron, advertising they are part of the solution. The message I take from that is they are admitting climate change is a big problem. So to me, that is the first mental leap they have already taken, whether that was mentioned in a positive or negative sense I cannot guess, I am not a psychologist or a psychiatrist. But I get excited when I see that ad. This is an oil company admitting climate change is a problem and they are part of the solution. So we have passed a major hurdle with them.

CS: Great stuff. You have been at this now for decades. I tried to learn about some of the major breakthroughs. From an academic standpoint, from a research standpoint, where are you taking this over the coming years? What would you like to accomplish at this point?

VR: If you look at the focus of Copenhagen and most of the climate committee, they are focusing on the long term. If we go on this route there is no doubt we are driving towards a precipice. But I am focusing more on the immediate future, 20 or 30 years and to me, my work suggests we are going to pass that two-degree threshold. For me that is already a precipice, and I am trying to extend that time. Instead of heading for two degrees in 30 years, I am trying to push that time to 60 or 70 years. That is my sole focus, simply because I feel we have to believe in the innovation of human beings.

CS: Thanks so much Dr. Ramanathan. It was a great honor to have met you.

For more information on this contributor, please visit:
http://2greenenergy.com/renewable-energy-facts-fantasies/.

NATIONAL RENEWABLE ENERGY LABORATORY

As one might expect, there are many non-profit organizations of different types that make important contributions to the quest for clean energy: government agencies, NGOs, trade associations, etc. The handful of such groups that have contributed to this book speak to the important work they are doing to develop key technologies and to adopt rational policy by which renewables can be moved forward in a concerted and responsible way.

We begin with National Renewable Energy Laboratory. NREL is the only federal laboratory dedicated to the research, development, commercialization and deployment of renewable energy and energy efficiency technologies. I was delighted to speak with spokesperson George Douglas.

Craig Shields: My readers and I would love to know more about NREL. Perhaps we could start with an introduction?

George Douglas: Sure. NREL began its life in 1977 as the Solar Energy Research Institute, or SERI, on the heels of the first set of OPEC oil embargoes. It was founded to explore how to convert energy from the sun into usable energy. That started to be more broadly translated fairly quickly to include wind energy and energy from biomass. During the first Bush Administration, SERI was elevated to national laboratory status and named the National Renewable Energy Laboratory, which is what it is today, to reflect the broader mission of the place.

CS: Wouldn't most Americans have said, "Well, technology is typically developed in the private sector?" What was the impetus—what was the thought process behind doing this in the public sector?

GD: Well, the first statement is not true. How did we get to the moon? Public sector development of technology. The Internet is public sector development of technology. Really, the model that people think about is the Bell Labs model—the long-term investment in technology. But after the Second World War business itself became much more increasingly interested in short-term returns.

And the role of government in investing in high risk and long-term research was given a great deal of credibility during the Second World War. The development of radar, development of nuclear arms, and so forth—specifically aimed at harnessing nuclear power. So Oak Ridge National Laboratory, Sandia National Laboratory, Los Alamos, etc. all grew out of that. So, there has been, at least for the last 60 to 70 years, the divide between what research is generally pursued by private enterprise and what research is pursued by the government and in academia. It's the difference between near-term results, and by near-term—I don't mean tomorrow—but in the 10 to 20 year time horizon, and much longer-term problems and results.

CS: Yes, well those are excellent points. Thanks for clearing that up for me.

GD: So, NREL was one of the later of the national laboratories, and our only mission is energy efficiency and renewable energy—although other national labs have energy efficiency and renewable energy as parts of their portfolio.

CS: Okay. Well here's another thing that typical Americans may think. And maybe we're wrong here too—and that is that the Bush

Administration 43 did pretty much nothing to forward an energy efficiency agenda, and now all of a sudden there is a great deal of renewed interest and activity in this. Is that true?

GD: Only partly. Through 2005, and early in 2006, the Bush administration's attention to energy started to include, in a much more robust way than previously, renewable energy and energy efficiency as well. In the State of the Union Address in 2005, President Bush talked about producing cars that run on hydrogen. In 2006 he went further and in the State of the Union Address talked about things that sounded like nonsense to me—among other things saying that we could make ethanol from switchgrass, which brought the great Jon Stewart diagram of a switchgrass processing device, which as the camera pulled away remarkably resembled the bong. It was kind of funny.

But also, President Bush visited this lab on Presidents' Day of that same year. At that point, just prior to that NREL had been faced with layoffs. For us that was sort of a turning point. President Bush came here, held a little panel discussion, a town-hall meeting style, and a great deal of publicity ensued. And what followed from that were a couple of specific things. We opened a new laboratory building and then the following year, got funding to equip the building. And have been on a track of rapid expansion since then.

CS: Great. In your estimation, what are your two or three most likely projects that will move the needle in a big way in the coming decades or so?

GD: One would certainly be cellulosic ethanol. We think that the science and engineering are very much on track for producing billions of gallons of cellulosic ethanol in the next 10 years. And that would be significant. The second thing that seems to be moving fairly rapidly is thin cell photovoltaics. The potential for thin cell in PV has long been that it can be manufactured less expensively, more easily installed, in a wider range of places, that sort of thing. The problem has been its efficiency, which remains lower than we'd like. It just takes a large investment to make plants that can produce huge sheets of semiconductor material. But progress has been pretty rapid on both fronts—increasing the efficiency with better technology on the manufacturing side. It seems like a lot of venture capital type money is flowing in that direction these days. So those two things particularly look very good.

CS: Excellent. You raise an interesting question when you say venture capital money, i.e., private. So you're suggesting that the private and the public sectors collaborate on projects like these. Tell me about that please.

GD: Yes, that's what we do all the time. NREL runs, for example, an incubator program. Federal money in there right now is in the tens of millions of dollars—for photovoltaics—in which promising companies get a boost in federal money, which provides them with both some money to live to fight another day, and also a credential. They've been reviewed by us, and found worthy.

CS: Like Solyndra?

GD: Solyndra is a good example. And so then those companies can take that credential and go raise private money.

CS: I understand. That makes a lot of sense.

GD: So we do that. We also have a number of what we call CRADAs, cooperative research and development agreements, with a variety of private companies, both big and small, on a variety of projects. We will give out three R&D 100 awards this year, which is not an unusual year for us. All of those awards are given for technologies—the 100 most promising technologies. And for all of those awards that are given, there has to be a product.

And of course NREL doesn't make products so there has to be an industry partner. Sky Fuels is one of our partners. We developed together a polymer coating for concentrating solar power—solar thermal electric—with great promise—it doesn't involve glass so it's less heavy, it's less expensive to make, it's easier to install, and so forth.

And so Sky Fuels is looking for a market. And I think they'll find it too.

These are the kinds of things we do at all kinds of levels of involvement with companies. We license our technologies. And like I said we work in CRADAs with companies, we do work for others where we might have facilities or equipment that would be very expensive to duplicate. But we can put together an agreement in which some of the information that comes out is proprietary and some of the information that comes out is public. Our interests tend to be in the basic science side, so if some basic science is being worked on—if there is some knowledge or discoveries we think has a benefit to all levels, and the company sees some benefit that specifically has to do with their proprietary process, then that makes a nice marriage.

CS: I just had a conversation with an extremely senior guy who was talking about biofuels and I said, "You know, all things being equal, wouldn't we rather be going in a direction that isn't trading in one form of hydrocarbons for another?" He thought I was an idiot to ask such a question. But, I mean, where did I go wrong there? If you had a choice between solar thermal, concentrating solar power, and switchgrass, wouldn't you prefer the former?

GD: Sure, except for the infrastructure. The infrastructure for personal transportation is all based on liquid fuels. And it's huge. Trillions of dollars worth of infrastructure. Changing that's not an overnight process. So, things are working in parallel. For the liquid fuels infrastructure, ethanol blends in very nicely. We know how to make flex fuel vehicles, we know how to make ethanol-only vehicles. I mean, all of those things blend in very nicely. However, at the same time, we are working on plug-in hybrid electric vehicles, which is a step towards all-electric vehicles.

Now, plug-in hybrids require some liquid fuels. And so, if the goal is to use less imported petroleum, which is the stated DoE goal, then a much quicker path to that is to start incorporating more and more homegrown fuel in the liquid fuel supply. For most personal transportation, you can probably get to electric at some point. But to go to electric right now, there is a range problem, and a charging infrastructure problem, and a refuel time problem. And those are insurmountable. I mean, scientists are optimists in general, and everybody thinks that we'll eventually have better batteries. The batteries today are better than they were 10 years ago, and 10 years from now they will be better still. But ethanol could be blended into the liquid fuel system pretty much right now.

Having said that, there are some problems. With ethanol you can't run it through a petroleum pipeline, for instance, so you still have a transportation problem. You'd either have to build dedicated pipelines or do what they do now, and ship it by rail or by truck. So there is a proximity problem, there is optimization of engines to better use ethanol; flex fuel is like the worst of both worlds in many ways. But those problems seem more quickly solved than the problems for all-electric vehicles. There is a vast agricultural market in this country that can grow everything we need to make ethanol.

CS: Well, you bring up an interesting point. And again, I'll go back to what my particular take on the typical American's viewpoint is, and in this case it's not very kind. I think that most people think that corn ethanol in particular is one of the biggest disasters, and in fact I'll go so far as to call it a malicious disaster, ever perpetrated on the American taxpayer. We could have known—or probably did know going in, that there is no advantage to doing this. By the time you end up irrigating and fertilizing and harvesting and processing uncountable of acres of corn, not to mention the fact that there's interruption of the food supply, you have a considerable negative net gain.

But I don't think it's a technological issue; I think it's perceived more as a political issue. In other words, if it weren't for extremely powerful senators in the corn-belt, this wouldn't have happened in a million years.

GD: Maybe, maybe not. The one thing to keep in mind, and again this isn't our argument, but one thing to keep in mind is that if the goal is to reduce the use of petroleum, then corn ethanol is a substitute for petroleum. If that's the goal, to reduce the use of petroleum, and thus reduce the use of imported petroleum which is really the goal, then corn ethanol does that. If the goal is to have a lot of cheap corn, corn ethanol doesn't help there either. But if the goal is to reduce the amount of petroleum you're importing for use in transportation, then corn ethanol definitely accomplishes that goal.

CS: Well, can you speak to the idea of reducing our dependency on foreign oil? If it was sincere, what about the first six years of the Bush administration? Why haven't we done more, if we are sincere about cutting our dependence on foreign oil?

GD: Well, you're asking me for a political argument. That really isn't something that a laboratory does. That said, let me say this. If you took a graph of this lab's budget and overlaid it with a graph of the price of a barrel of petroleum, you would see that the higher the price of oil, the more money the laboratory's got. The lab got a lot of funding during the Carter administration when oil was expensive. In the Reagan administration, particularly in the latter part of the Reagan administration, when the Soviet Union started pumping lots of oil into the US and the OPEC cartel fell and we ultimately had among the lowest gas prices ever was when our severest cuts were.

CS: OK. So you're suggesting that there is in fact pretty decent sincerity. You're suggesting that there is no reason to be suspicious of the sincerity of this claim.

GD: I'm saying that the sincerity of the claim has had, I won't speak for now, but has had more to do with the price of petroleum then the geopolitical situation. That seemed to change somewhat after 9/11. Now the concern may be higher on the national security side than on the economics side. Or starting to rise, almost, not higher than, but starting to.

CS: Well, do you know James Woolsey?

GD: Right, I do know James Woolsey.

CS: Well you know that he puts forward seven or eight reasons for moving away from fossil fuels—each pretty much independent of the rest. And I see his point. As a loyal, tax-paying American, I don't understand why this is at all controversial.

GD: Well, there's no political vote for suffering. I mean, the real way to reduce consumption of petroleum is to raise the price. We know that. I mean, when the price went up to five dollars a gallon a year and a half ago, consumption immediately dropped. Immediately. So, if the government was to say we are going to reduce the consumption of petroleum, slap a three dollar tax on it and that would do. Now, you find a congressman who's willing to propose that.

So many people find merit in your argument. But the political will to get there is a more difficult thing. What we want, or what seems to be wanted, is this progress without pain. And technical solutions sometimes offer that.

CS: Well, let me ask you this. If you're big energy, say Exxon Mobil or Chevron, you don't want renewable energy to happen and you spend a lot of cycles both hidden and obvious to throw monkey wrenches into the works of clean energy. An example might be Exxon saying, "OK, we won't fund sham research projects that try to discredit global warming."

GD: I've seen the kind of evidence that you are citing, but from NREL's point of view, what we see is increasing interest and cooperation from large corporations including big energy corporations. We have CRADAs with Chevron, we have CRADAs with BP. In addition to them, Dow and 3M—these are corporations who I think are beginning at least to see themselves as energy companies, not oil companies. Chevron is involved with us in a CRADA researching algae oils. So the potential there is enormous. And that you could make an oil, a petroleum substitute, from algae. Actually, algae makes the oil; you just harvest it.

And the Chevron side of this whole thing is the processing of it. We can grow algae that makes oil. Not a lot, but we can do it. But once you got the algae oil, you've got to turn it into a product. And Chevron's very good at turning oil into products. And so, that's their interest. The fact that Chevron's money is in it—it's not a lot, but several million dollars—is an indication that they're interested.

CS: Well, certainly if somebody's going to do it, and I guess Chevron's board of directors said "Look, somebody's going to do this so why shouldn't it be us?"

GD: I think that lots of things have moved forward in the last 10 years that the general public perception is probably behind the curve on, because it takes a long time for old truths to become new truths. And so there's long been this sort of oil conspiracy theory, that there is some guy with a garage in Arizona who can make rocket ships out of water, all that sort of stuff, which a certain group of people, I suppose, believe.

But there is also this sort of notion that progress is being made more slowly than it should be. And that may be more slowly than it should be may be true, but it's not more slowly than most technology curves. So what people don't tend to recognize is that they've got an iPhone now because it came from the development of technology. They think that just suddenly, here is a phone that will do all these amazing things on the marketplace, but the basis of that technology goes back 60 or 70 years. And the curve gets more rapid as things develop. Sort of like Moore's Law.

But in renewable energy, we're not at that point in the curve. But that point will arrive. I mean, you can look at the history of lots and lots of technologies including what we would call ancient technology like steam engines and the curve is the same. When you consider the discovery and its usefulness over time, it's really incredibly similar. And so, no one has any reason to believe that, for example, the efficiency of photovoltaics, the ability to put photovoltaics on flexible and different materials, the cost of photovoltaics, won't follow a very similar curve to other technologies. Same with cellulosic ethanol.

CS: Great analysis. Thanks. And I guess what readers want to know, just to end off on this political point of view, I guess readers just want to know is this happening in good faith? In other words, if this is slower than everybody wants, but everybody's trying and nobody is deliberately sabotaging this effort, just because they have an interest in doing so, that's fine. But here's the problem: there have been examples of that in American commerce. Efforts to sabotage progress surface all the time. But I think people want to know, is there a real sincere effort that is being essentially undeterred by powerful corporations who have extremely clear-cut missions to sabotage it?

GD: Well, what I'm saying is that we work with, have worked with, continue to work with, and at an increasing rate, with traditional energy companies.

CS: OK, well that's good to know.

GD: You know, things gather up steam and at some point, and I'm not trying to say that this is all what happened, but at some point it may look like the development of a new technology or a different political look at things is bad for your business. And so you resist it. And then there comes a point when you've lost that game. Things are shifting. And they're beyond your control. And smart businesses figure out a way to change from making leather saddles to making leather seats for cars.

CS: Yes, exactly. If you can't beat them, join them, in other words.

GD: Yes. Figure out how your business model can match into that business model. And I think we're seeing an awful lot of that. I read comments almost every day where somebody says we believe a carbon penalty is coming. It looks like it might be cap and trade, could be a tax, whatever. The guy who is making electricity from coal is going to have to deal with it. And the guy who makes gasoline—he's going to have to deal with it. And there are all sorts of ways to deal with it. You can raise your prices, you find a cheaper way to make your product, you can switch technologies. I mean, it's just constant that you see that.

Well, five years ago, oh no. A carbon penalty will cripple the economy. Can't have such a thing. Now, you still hear that from some people, but most people in business, what you hear is it's inevitable. We are coping. So there's been a shift.

CS: Yep, I'm with you. Now let me ask you one final question, if I may. Every time I hear about renewable energy, the idea that we need a mix seems to be a foregone conclusion. We have wind in the plains, we have solar thermal in the desert, we've got geothermal in the mountains, and all of these things are a good thing and they add up. You don't want all your eggs in one basket, etc. I'm wondering, if I were the king of the world, what's the matter with putting a huge set of solar thermal farms in the otherwise completely worthless land in the southwestern deserts and building out the grid with high-voltage DC and storing the energy, as required anyway, with molten salt? What's the matter with just studying the problem and picking the best solution?

GD: Well, a couple of things. First, I'd advise you not to go to Barstow and say that the desert is worthless.

CS: Well, that's a good point. I have to stop being that cavalier.

GD: I'm not sure that there is absolutely enough resource, I don't know if water is an issue or not. NREL doesn't think that that's the wisest course for a couple of reasons, though it could be done. One is that it would require a great deal of super transmission, which would cost a great deal of money. So somebody has to say that they are willing to invest in a 20-year transmission build out. And that's, well, in our view that should be done anyhow.

CS: Yes, well that's what I was going to say. That probably needs to be done anyhow.

GD: Committing to one technology... I don't know. I suppose that your idea is not necessarily a bad one, but we think a better one is diverse resource. There are security reasons for that. There is also the notion of locally generated power that generates local jobs and a variety of other things. So there's some economic considerations. We think it would be less expensive overall.

CS: Oh, okay. I would've thought the opposite, by the way; I would've thought that the concept of scale would have suggested that choosing one as opposed to many was, all things being equal, a good idea.

GD: I think that is valid to a point, but not after it. I think the economics of locally produced and used power reduces need for storage and other very expensive portions of the power grid. I mean, storage right now is extraordinarily expensive. So, you know, storage could be cheap in the future but right now it's very expensive.

CS: Right. Well, I think the thing about molten salt I like is that it really does have the potential to change that equation and especially if you're talking about solar thermal which is already heat energy in the first place.

GD: Right. We are very fond of CSP (concentrating solar power) with storage. There is a water issue there. And it's not nearly as intensive as agriculture, but it is more intensive than, say, a coal-fired power plant. And the desert Southwest is a place where there's not as much water as you'd like there to be

Now there's technology being worked on that... there's air cooling and other kinds of technology that's being worked on to reduce the amount of water used, but again water is the cheapest way to cool right now. There was an article in Scientific American a couple years ago that was very big on using lots and lots and lots of CSP as a wedge, a climate wedge. But generally speaking, we are, as a laboratory, not sold on any current technology as a 100%, or even 10%, solution. A couple of things happen if you commit to a specific technology. One is that you stop exploring other ideas and such.

CS: Well that's a good point.

GD: Now, if I were a utility, say, and I needed to make a lot of carbon-free electricity, and I could get a hunk of land in the desert southwest that was the size of Dallas Air Force Base and start building them there and had all the capital I could come up with and could convince the government to build a lot of transmission out of there, I might do that.

Because with, I don't know, 10 years of concerted investment, I could start making money and make a lot of it. Just like building a coal-fired power plant—or a nuclear plant, as a better example really, because the risk is that high. But as a research institution, our notion is that all of these things can have their place and there may not be any clear winner because the technology applies best where it's used best.

CS: Yes. And there are, as you pointed out, individual local political and economic concerns that may trump the idea; you can't have a tens of millions of people out of work. You have to figure out what you're going to do with all the people who used to be mining coal, as an example.

GD: Right, there are those kinds of political and policy concerns, but beyond that there is the elegance of making and using energy as locally as possible. Which as, obviously has the economic aspect of employing people there and so forth. But also, you know, unties you from a grid system that can easily be brought down.

CS: Yes, if I were doing this by the way, it wouldn't be one big square. That's for sure.

GD: We use that "hundred by hundred square" largely with PV, but it works with CSP too, as a demonstration of the rather small amount of land area in the US. But people always misunderstand it. I've almost stopped using it. Because they think that's what you want to do.

CS: Right. Now, you've mentioned the oil and national security issue, but you haven't talked too much about global climate change. What's your position on this?

GD: I'm concerned that the amount of carbon in the atmosphere is going to affect the climate, yes. And that concern is of course shared by nearly all today.

CS: You say "nearly all," yet about a year ago it seemed like it was "damned near all." It sounds like there's been a little bit of backpedaling. Can you explain that?

GD: Economics, I think. That's my opinion anyway. I think what happens is everybody says we need to do something about the amount of carbon we're putting into the atmosphere. Then the things that we need to do are starting to get some pushback. And I don't think it's a bad thing—actually—the pushback. You get some pushback that tends to be based in "can we afford it?"

It's not necessarily a bad thing because it makes people go back and look at the cost of the technology. You know, you can get all starry-eyed about putting PV on every roof in America, but when you look at the costs of putting PV on every roof in America then you've got to wonder where the money will come from. And NREL was the lead author on a study called "30 year wind study." It's 20% wind by 2030. And study asked, "Is this possible?" And the answer to that was Yes. And then it asked, "Well what are the challenges to getting there?" And a big challenge is just building enough manufacturing facilities.

So the biggest challenge's transmission, but one of the top five was just building enough places to make wind turbines. And that would be the same of any technology you chose.

CS: Oh I'm sure of that. The thing about solar thermal, when you think about it, you try to envision a square of that stuff 110 miles on a side, that's a lot of concrete and aluminum and glass.

GD: That's right. So it's a huge endeavor no matter what you do. And then the other thing that I think many people sort of lose sight of is that the energy system we have today, in the developed world not just in the US but in the entire developed world, is incredibly successful. We lose, I know that the banking industry would like it to be nine nines, you know, 99.999…%, but it's like five nines right now. And in math for any system to work that well is pretty amazing.

And then, the cost. The average US electricity cost is something like nine cents a kilowatt-hour. It's nothing. And to think that this is the system that we are trying to somehow replace is daunting. Because it's not like taking something that's terribly broken and fixing it.

CS: Well, I would say that it's broken, but most people don't see how broken it is. Most people don't figure that their father who just died of cancer probably died because of something associated with the way we generate energy or grow our food or both.

GD: Yeah, all of those externalities are prices we've long paid. I mean, not just with the electricity system. I mean, our life expectancy was a lot less then... 50 years or 100 years ago before we had an electric grid than it is today. I mean, the environmental consequences of being alive have long been harsh. And they're less harsh today and electricity is part of the reason why.

And so, even though system is not perfect, it is so by far... the fact that I can go in and get my heart operated on, or my lungs, or whatever it was that you were thinking of causing this, is due to electricity.

CS: That's absolutely right. That's a very good point.

GD: So these kinds of things, I guess what my point is, the point I'm trying to make, is this is a system that works very well. It's not all perfect, but it works very well—and very cheaply, for most people. And there's a great deal of resistance to making it work less well and more expensively in the name of some risk that's in the far future and intangible.

CS: It's intangible—and it's coming under fire too. In other words, there are people who believe that global warming is a hoax. I'm not one of them, but you hear that all the time.

GD: Yes, there are. Most people just aren't thoughtful about it. They just don't pay attention. And why would they? I do because it's my job, but I don't pay attention to a lot of other things.

CS: Well this has been fascinating. You don't happen to know the name Peter Lilienthal (ex-NREL researcher), do you?

GD: Yeah, I know Peter.

CS: Oh, good. I've been speaking with him on this hybrid power management thing.

GD: He's got a lot to talk about there.

CS: Oh yeah, he's amazing; he's a wonderful guy. And the reason

I bring him up is because he was the first guy who really started going through this cost argument with me when I started talking about the ideal way of handling this. He goes, "Look, there's plenty of renewable energy out there. The question is what you want to pay for it?"

GD: That's right. And that's the real question that NREL faces. Now keep in mind that NREL does a lot of different things, and one of the things that we do is very basic scientific research. And we have people who work very much in the depths of material sciences and quantum physics and those sorts of things. But a larger task for NREL is trying to develop technologies that we can help move into the market place fairly quickly.

We have a nano-science guy who looked at the behavior of the atomic structure of silicon at the nanoscale, and can get, in rare circumstances—this isn't it at all ready for prime time—but they can get two, and I think even three, what they call excitons, when a photon of light hits the silicon atom. So, you know, a photon hits it and normally one exciton comes out. And the exciton is electricity.

CS: Yes, I've heard this.

GD: Okay, so, does this have the potential to double the amount of electricity you can get? Sure it does. Triple, even. But, can it be done in any way other than under an electron microscope with all kinds of special circumstances? Not now.

CS: Right, I'm with you.

GD: So, these are the kinds of things that I think the deep science can bring us.

CS: Fantastic, Thanks so much. This has been of enormous value, and I'm really grateful.

GD: Please do stay in touch. Good luck with your book.

For more information on this contributor, please visit:
http://2greenenergy.com/renewable-energy-facts-fantasies/.

RENEWABLES AND ENVIRONMENTAL STEWARDSHIP— THINKING THROUGH ALL THE IMPLICATIONS

Renewable energy has numerous obvious advantages over burning coal and other traditional power sources. Yet each of the various clean energy technologies is accompanied by a certain environmental impact, all of which need to be understood clearly.

Audubon's mission is to conserve and restore natural ecosystems, focusing on birds, other wildlife, and their habitats for the benefit of humanity and the earth's biological diversity. Brian Rutledge functions as Audubon Wyoming's Executive Director, and provided this very candid and poignant conversation.

Craig Shields: I've read some of your stuff, and it's excellent. We want renewables, but they come at a cost—and part of that cost, ironically, is ecological. Is that what you would say?

Brian Rutledge: Exactly.

CS: So, as a person who's actually studied the issues, what are the costs associated with migration to renewables?

BR: Well, first, I think there's a huge opportunity here. We do need renewables to work. But I'm not utterly convinced that we've done our homework on the technology. I think we're racing quite a bit right now.

We have good intentions, but we need to be deliberate in all our moves.

We've seen before, when we tried to develop the supposedly clean natural gas, what that's done to the environment. We've also had the early experience with the renewables with damming all of our rivers. We're still in the process of trying to recover from that. So to me it seems like there's a great deal to be studied here, and then we need to do this with care and caution—rather than racing in and grabbing at the first opportunities, saying to hell with the environment and the animals and the wilderness.

CS: I don't think anyone could dispute what you're saying, but can you give me an example?

BR: Here's what we've seen in our direct work in trying to cope with the greater sage grouse—a species that is walking on the brink of extinction—and the experience around it with renewable development. The bird was somewhere about 85% and 90% greater in numbers than it is now at the turn of the prior century. If you read about it a little, you'll find that George Bird Grinnell and Teddy Roosevelt camped at the foot of some bluffs south of Casper and when the birds started flying down to graze along the river, they blacked out the sky until noon.

And now we're down to a couple of hundred thousand birds. We started off by first fragmenting the sagebrush habitat, then going through a long process of where the belief system was that the only good sage brush is dead sage. So you obliterate sagebrush by fire.

And I want to stress that this greater sage grouse is a litmus species— one that tells us what's going on with that whole ecosystem. It's not a single system management question; we look at the sage grouse as an indicator as to how the rest of the system is doing. There're at least three other species of birds out there now that are in even greater trouble than the sage grouse, they're just harder to study.

CS: I see.

BR: Sage grouse all show up on a "lek" or dancing ground in the springtime and you can count all the males, and if you work hard at it you can count quite a few of the females. And so you'll know how the population is doing year by year by doing these lek counts. So now we've got years of lek count studies and then we've found that as coal bed methane began to be developed, largely in the Powder River Basin, and as coal or as natural gas methane began to be developed in deeper seams in western Wyoming, we've seen massive fragmentation of the land and plummeting decline in the numbers of birds. In Powder River Basin, declines are in the neighborhood of 84%. So areas that were once gems of great sagebrush are no longer.

We had negotiated very hard in Wyoming, I will say, for two years to build a system by which the gas industry, agriculture, the conservation community, and all the agencies negotiated a deal that would protect about 80% of the birds in the core areas. We established a deal where the perimeters of those breeding areas were led by peer-reviewed research, and we decided on a methodology under which those areas could be developed for gas—but in such a way that they wouldn't diminish the number of birds. And that is a 5% cap on development per section. So each square mile would get a maximum development of one drilling pad and all roads, pipelines, power lines, all the rest of it, would add up to no more than 5% of the entire section. We all agreed to this, and began working on it.

But then suddenly the wind boom started, the wind people didn't figure that any of this applied to them. And we tried to have a negotiated discussion with them and were basically told, "We're saving the world, we don't need your permission; we're going to do what we want." And it was really astounding to have people doing this kind of capitalist charge in the name of the environment.

If there had been any real thought and concern, this would have been done in southeastern Wyoming, because southeastern Wyoming already has a human footprint. And we've done a lot of the groundbreaking work on that—on establishing where the renewable resources are available in well established human footprints. And in the case of Wyoming that turns out to be the southeastern quadrant. But the wind people want to do it out in the sagebrush. They say, "We need to be closer to Vegas;, we need to be closer to shipping our fuel, our product, south and west."

Well, guys, you know that's gonna destroy this species and this ecosystem. And they reply, "Not our problem." And besides, and this is my favorite, "You don't know how a wind farm's going to affect it." In other words, the development of a wind farm may have a different impact on the grouse than development of the gas fields.

CS: Well isn't that possible?

BR: Yes, it's quite possible that it would be a much greater impact than the gas fields.

CS: Well, ok. But I mean it is true that nobody knows?

BR: No, because we know from research how the birds respond to vertical structure on the landscape. Now, I don't know how you do a wind farm without vertical structure, but if they can show me I'm very interested.

CS: Well let me ask you this. Philosophically, I can see where you're coming from on this. And by the way, I'm not trying to be antagonistic, but my job is to ask the tough questions. Is it ok if I fire one at you?

BR: Absolutely. You won't be the first.

CS: Ha! I can see that. Part of me says, in a comparison with coal, there's absolutely nothing that you could do short of a deliberate nuclear holocaust that would be worse for the 6.8 billion people on this planet than to continue to burn coal. So if it really is a trade-off between a couple of species of birds and the incredible near-term damage to every living thing on this planet, not to mention the long-term environmental damage, then maybe we do have to make a tough decision.

BR: But let's make it a real trade-off first. Let's understand the parameters. And when we look at them we don't see that as being necessary. Look at Wyoming. Realize it has sixteen times the wind resources that are needed to develop the amount of wind energy suggested to be sort of our share of what's needed nationally, and you realize that there's plenty of opportunity within the already developed footprint to do this. Why in the world would our first decision be to destroy a challenged ecosystem? If you've built everything else out and can't go anywhere else, in that case we've got a real argument.

CS: Right. That part I understand for sure. The part I don't understand about what you just said is the thing about "sixteen times." I would say that we're counting on Wyoming for more than its share per capita or per square mile of land. Wyoming has wind, just like deserts have sun, mountainous regions have geothermal, etc.

BR: No, I'm talking about the way they've doled out the amounts expected from various areas. We haven't even built the resources in to take it out of here. The real problem is transmission. Guys are talking about building power lines, which are going to have their own sets of difficulties, all the way from these fields in Wyoming to the end buyers. This is all about coordination or EIS (environmental impact study) work.

CS: Well, I also have been speaking to the solar-thermal people, who of course envision a few thousand square miles of solar-thermal farms in the southwestern deserts. And it is true, I think anyway, that you're going to be messing with fewer species, because you're talking desert there. But at the same time you're certainly going to have to transmit the power out of there. You're going to be doing some kind of grid build-out that includes high voltage DC.

BR: Well I wish they'd talk more about that. Because they keep talking AC. And it's hung AC. It's not buried DC. It's hanging this maximum 500kV lines fifteen miles apart because of some very old regulations. And it's using new corridors in so many cases, instead of multiple use of both corridors. And so there's an awful lot of discussion that needs to occur here.

And I'm not the expert on all this. But I do know what I've worked on with the Western Governors' Association and what I've worked on in Wyoming, and I can assure you that this has not been studied, planned, or delivered in any way that we can have confidence that we're ready to roll with it. I keep hearing terms like "smart grid" and then I look at what's planned for wind and every wind zone or hub is to be backed up with a fossil fuel plant? How is that smart?

You know, the grid should be taking those electrons as they come and doing it smart and taking it from fossil when it has to, but taking it from renewables when possible. We're not even talking about segregating lines. The Wilderness Society has told me how concerned they are about building new transmission for wind up in the eastern Wyoming. And that revitalizes, this is the potential not a reality, but revitalizes the five coal fired plants that were planned for up there at one point. That's just wrong.

And we have not planned this. I have a real problem with the way we've done the stimulus funding, because we're telling these guys you have to have a shovel in the ground by next fall or no stimulus money. It's like an Oklahoma land rush. It's the wind rush. And we're screwing ourselves by doing it. There's no doubt in my mind that we are setting up some just horrible errors. And so for me to say "Ok, yeah, I should step back and go pro-wind and to hell with the sagebrush," I can't do it.

CS: No, I'm with you. Well I am particularly appalled at the idea that these things have to have co-located coal-fired power plants as backup. Are you sure about this?

BR: Well, this is what we were told. This is what we were doing. The REZ, the Renewable Energy Zone, the QRAs, the Qualified Resource Areas, and all this work with the Environment and Lands Committee and the Western Governors' Association.

There were wind power folks on all these committees, and they insisted that we build hubs instead of renewable energy zones because we were able to get out of the guessing game of what might be there in the way of resources other than wind, other than solar, other than geothermal, so that we'd stay out of that battle and get something done. We set up hubs because they said we have to be within a fifty-mile radius of the hub. And as soon as we said that, they said "Yes, and we have to have a place in that hub for a backup plant for when the wind's not blowing, so we can have a fossil fuel fired backup generator." That's not a smart grid.

CS: You can say that again.

BR: Now the really bizarre thing about coal, and I couldn't agree with you more about the burning of it, don't take me wrong, but the thing about coal is we built protections into their systems, because they were considered "maximal surface disturbance." We built forced systems requirements for pre-planning—SMCRA (Surface Mining Control and Reclamation Act)—a federal regulation that controls the way coal is developed. First you have to have a plan for reclamation before you can even start to dig. You can't do anything without having it figured out: what you do about the species that are there, what you do about the land that is there, and they have become pretty darn dependable. They've got a regular culture, including permanent staff, that works on making sure they're doing their reclamation and improving habitats and all those things that are required. On the other hand, the cleaner fuels, both wind and gas, do not operate under SMCRA because they are considered minimal surface disturbance.

Now you only have to see a wind farm to know that that's not true—but nonetheless they are considered minimal. So now we've got federally approved safety roads all through these gas fields, so you know in many cases sixty to seventy foot wide road beds that go everywhere in these fields and some of them are ten acre developments. So it's massive surface disturbance. But because of the way the regulation is done, they're exempted. You find no requirements that they do anything about the environment they're damaging. So though coal does the greater harm in the air, they're much easier to get along with on the ground.

And wind doesn't figure it falls into any of this. We're working with technology that was developed in 1942 and is virtually unchanged. I just have to believe we can do our homework better on that one.

CS: Yes, well I see where you're coming from. Now I often make the point in my blog posts that we as a nation seem wedded to this idea of five or six or eight different sources of power. And it seems to me...I live on a horse farm here in California, and if I were going to take this place off the grid, I don't think I'd be having geothermal holes dug and wind turbines in the pastures and solar panels over on top of the barn, I'd just pick the best one and do it. I don't see the imperative to have all these things balanced at the same time anyway.

BR: I am with you. I am very much in favor of renewables. But I did hear something really interesting John Stewart said on the Daily Show night before last, talking to Al Gore. "Tell us the one thing we need to do." And I know that's oversimplifying, because nobody really can tell you the one thing, but we sure do need to get our requisites in order.

CS: Right. The other thing I wondered, based on talks I've had with other folks, is this: does big oil and big coal protect its turf with very cozy relationships with lawmakers?

BR: Yes, they have the biggest lobby in the world.

CS: Yes. You must see a lot more of this than I do from my desk here in California. If you were king of the world, are there things that you would make sure never happened again?

BR: Well, first of all, we would see that we never had a 2005 Energy Act again. This is where we said that energy development is paramount over all other interests on all public lands. We've still not overcome that. It created a culture within our managing agencies, and it's been devastating for them. We would never again have Dick Cheney or his offices calling field biologists in their stations around Wyoming and telling them which wells would be permitted to drill regardless of what the issues might be on that well.

And that happened every other week. Never again would we have some of the mining misbehaviors we had—I'm sure you saw the cases of sexual misconduct amongst the agency folks in dealing with the mining group out here—Lands and Minerals. We would not have nearly the relationship between any of the federal employees and these engaged industries that we have, where we have industry deciding what reclamation looks like, and industry deciding what impacts are acceptable. We've created a really sorry situation. And we can't really blame the staff when they've been under this for so long that they don't know who they work for anymore.

You may have seen the article on categorical exclusions: "Is the BLM Practicing Safe CX?" Playing on safe sex.

CS: CX –categorical exclusion? Could you explain that, please?

BR: Right. This was especially true in the gas fields. If you saw an area you wanted to drill, and if you did it on a minimal basis, you'd get what was called a categorical exclusion from doing an environmental impact study. You didn't really have to do anything, you just walk in and drilled your well experimentally and saw whether or not you got gas. And so they did a lot of this. And what they would do is build out a field, planning ahead, they didn't tell anybody this and I'm maybe being presumptive in saying this is exactly how they're done, but they drill as many wells as they could get done while under categorical exclusion. And only when almost all the damages were done, they would then file for an EIS to do full field development. And so you had the effect of a full field, avoiding all the environmental work by doing categorical exclusion.

CS: I see. Let me ask you this. The renewables people—in this case wind—pretty-much exclusively—have disappointed you. Are you saying that they are truly worse than the fossil fuel people, or that you expected better from people who ostensibly wanted a cleaner, better world? In other words, it's not that they behave worse—it's just that you expected better.

BR: Yes. You know, I told someone the other day—and I'd pretty well be prepared to defend this—the wind guys have acted like the gas guys on steroids. It's really unfortunate. I think that's forced though, by what I talked about with the supplemental funds that are based around getting the shovel in the ground now. We've set up a land rush. What did we expect?

CS: Wow. I knew that this was going to be a tearjerker, and I was right. It's a very sad story. And you must live it every day.

BR: Oh, more than I care to. Take this away from our discussion: my favorite word of late has been deliberative. I want to see a deliberated approach. Don't just rush out there and start slapping up a wind farm. All you have to do is start looking at the structure here including, which everybody fails to do, the footings in the ground. They're talking about a cement plant in Cheyenne—a new cement plant, which will be a new polluter in and of itself, just to be able to put in the footings for these wind farms. We need to be looking at all of this.

You know, we were crying about honeybees a year ago. We're losing them on a planetary basis at an incredible rate.

We need to pay attention to bats too. We're incredibly dependent on some bats—far more than we realize. And these wind farms just wreak havoc on bat life. They suck their lungs right off their face. I mean, the work on that is enough to just break your heart—especially if you're a fan of living things.

CS: I'm with you. Brian, thank you so much and keep up the good fight.

BR: To be sure.

"...especially if you're a fan of living things"—a clever irony, I thought.

For more information on this contributor, please visit:
http://2greenenergy.com/renewable-energy-facts-fantasies/.

THE CARBON WAR ROOM AND FREE MARKET CAPITALISM

Founded by British industrialist and humanitarian Richard Branson, the Carbon War Room uses the principles of laissez-faire capitalism to deal with the problems brought about by our planet's addiction to fossil fuels. The organization points out, "Our global industrial and energy systems are built on carbon-based technologies and unsustainable resource demands that threaten to destroy our society and our planet. Massive loss of wealth, expanding poverty and suffering, disastrous climate change, water scarcity, and deforestation are the end results of this broken system. This business-as-usual system represents the greatest threat to the security and prosperity of humanity—a threat that transcends race, ethnicity, national borders, and ideology."

I have known the group's CEO, Jigar Shah, for a few years now, since his leadership role at renewable energy pioneer SunEdison. I tracked him down at his office in Washington, D.C.

Craig Shields: Why don't we talk about the mission statement first. What are you trying to accomplish here, and how are you going about it?

Jigar Shah: Well, what the Carbon War Room believes is that we need to harness the power of entrepreneurs to promote market-driven solutions to climate change. And the reason for that is because we don't believe that global warming is entirely an environmental problem. Deforestation in tropical rain forests and those kinds of things are environmental problems, but carbon dioxide is not an environmental problem. Carbon dioxide is a very natural molecule that we all exhale. So you can't solve it by saying that it's a toxic waste problem. You have to solve it by asking, "What, fundamentally, within our method of economic growth and capitalism needs to be tweaked to be able to set us on a course to do business in greater synergy with our planet?"

And that's not just putting a price on carbon. The Europeans have just put a price on carbon and it hasn't worked. They've been charging $7 a gallon for gasoline since the 1970s, and they are not all driving electric vehicles.

So, it's more than just putting a price on carbon. I'm not suggesting that putting a price on carbon is a bad thing, but there is more to it than that, and that "more" is what we are looking for.

CS: All right. Well, give me an example of that. I'm an entrepreneur... how might your activities affect my activities and behavior?

JS: So, in the shipping industry, there are many entrepreneurs, we've counted 43 but I'm sure there're hundreds more, who have technologies that reduce the use of bunker fuel that ships burn: low friction paints on the side of ships, propeller designs, hull lubrication, etc. right? So why, since 1970, have almost none of these technologies been widely adopted except for maybe being pilot-tested?

CS: I don't know.

JS: Because the ship owners don't pay for bunker fuel. It's a pass-through to the shipping customers. So why don't shipping customers demand it since they are paying for it? Because they don't actually know what the miles per gallon is of all the ships. They just assume they're all the same. And so we partnered with RightShip to publish data that showed the miles per gallon of the ships... we had to collect it and then publish it... and now 160 companies from around the world are choosing the most efficient ships and thereby saving themselves millions of dollars a year worth of fuel savings.

CS: That's amazing.

JS: Right. That's a market failure that we plugged. I mean, it sounds simple except for the extraordinary amount of work on our part, but it sounded simple. But it will lead to 400,000,000 pounds of CO_2 equivalent emission reduction from black soot, NOX, SOX, and CO_2.

CS: Incredible. Now, I would've thought that that would've been a rare example—if not unique—insofar as people who are paying fees for things normally know what those fees cover. Do you think that this is representative of a class of situations in which the market really doesn't know what it's doing?

JS: The problem is about focus. If you're Wal-Mart and you believe that you've got $12 billion in your capex (capital expenditure) budget per year, why would you spend a dollar of it on energy efficiency even if I can prove to you you're going to make a 20% rate of return on investment? You'd rather spend that money on buying up a whole chain of stores in Spain and expanding your international reach.

That's your core competency. And so the problem with energy efficiency is if, as a homeowner, you have a credit card, why would you spend the money that you have left on your credit card to buy new windows for your house even if it's a 20% rate of return when you would rather spend that money on a vacation for yourself to Turkey? And so the problem is that you are giving people choices with their money and they choose not to use less electricity.

You have to figure out a way to take away their choice—and there are two ways to do that. One is mandates, but the other way to do it is the way that (solar energy pioneer) SunEdison did it, which is to say to people, "Hey, I have this free pot of money for you, but the pot is only for solar energy. Would you like me to put solar on your roof?" So I'm not asking you to use your credit card or the money in your bank account for this. I am saying to you, "Would you like to opt in to my money to upgrade your facility?" Then they have no choice. I mean, what rational person would say, "No, I don't want to use your money to upgrade my house." And then of course I charge you a surcharge on your property tax bill, but you're saving two dollars on your electricity bills for one dollar your property tax went up.

CS: Yes, that was an amazing business model indeed.

JS: Right, and that's what we did at SunEdison, and that's what we need to do in energy efficiency: find and correct market failures. You could solve the problem by passing a law mandating everybody hits some Energy Star rating for their building and just forcing it on them. But you could also take away their choice by saying this is a proposition that's too good to be true. Why would you possibly walk away from this and not opt in to this money?

CS: Great. So, tell me a little bit about the players. That's a pretty impressive cast of characters there.

JS: Well, you know, the Carbon War Room was started by Sir Richard Branson. His sense is that entrepreneurs are really the ones who solve problems. Since 1980 in the US economy, the Fortune 1000 has shed 5 million jobs in the US, whereas the economy has actually created 34 million jobs. 39 million jobs were created by mostly businesses less than 5 years old—entrepreneurs. So the real question is how do you get these entrepreneurs engaged in a way that they are going to want to create jobs in this area? How do you create that enthusiasm? And there's nobody more attuned to this than Richard Branson, the world's most famous entrepreneur. He has started over 300 companies, most of which are profitable and doing well. And so then he asked, "Are there other entrepreneurs that I can bring on board that are really amazing that could lend their expertise and resources to this mission?" So he brought in folks like Craig Cogut from Pegasus Capital, Strive from and Boudewijn Poelmann from the Dutch Postcode Lottery. We have many wonderful people from around the world these guys are just amazing entrepreneurs.

Entrepreneurs think differently. They don't say, "Well, we need to have all of this stuff lined up into small buckets—we are going to need this firm feed-in tariff in Germany to make the numbers work before we actually decide to start a business." They say, "Hey, there looks like there's an opportunity there and I'm willing to scrape and scratch and innovate my way into realizing that opportunity."

Richard Branson asked, "Well, what's holding all those people back?" There are clearly hundreds of people who are already starting solar companies, etc., but in all the areas of climate change, what's really holding those people back? And in many cases the market just isn't inviting enough to allow those people with their precious capital and time to start a business in this area. So, as a nonprofit, how do we make the market more inviting to allow those people to realize their goals of creating climate wealth?

CS: Wow. Is there a standard answer to the question: How do you make the market more inviting? I mean, I'm trying to interpolate between the examples you've given, but...

JS: Well, there's no standard answer. You start by realizing that technology is not the bottleneck, for at least 50% of global emissions today's technology is cheaper than the polluting one. If we can show that it's economically better today then policy is probably also not the bottleneck, because if you have an economically superior solution generally, the policy is not holding you back. You must have a market failure that is holding you back. Market failures come in all types from information gaps to transaction costs to market power.

There are many standard market failures that we study in business school. We identify them and then figure out which resources we have to bring to bear to eliminate them.

CS: So does the public sector play any role here? And if so, do you run into people's native distrust of government?—"I'm from the government, I'm here to help you?"

JS: I think the government's problem is that in a democracy the government can't put down an authoritarian response. And so, in a democracy as vibrant as the United States or India it's very difficult to make sweeping changes to industrial policy and energy policy to provide all the right market signals to move everybody in the same direction.

CS: I agree, but I think that if you interviewed the hundred CEOs of the Fortune 100, pretty much everyone would agree that we are headed toward a big fat tax on carbon.

JS: I don't think that's true. I mean, today, I would say that everybody uses the right talking points about a "big fat tax on carbon," but if you look at the activities behind the scenes, we had a perfectly good bill that has basically rewarded—and is transferring wealth to—the coal industry and other industries in exchange for their votes to get the bill passed. Yet there is not a person that I know in the United States who thinks that the climate bill is actually going to pass. So either these CEOs are just not powerful enough to move this across the line or they're saying one thing and doing another.

CS: Do you see corruption in this arena? For example, some say that we have corn ethanol because of a few powerful senators in the corn-belt.

JS: I don't agree with that. I think that the challenge is really that, after technology and policy have been shown to meet their hurdles, that there are market failures in the way of scaling up the solution. The problem that we have with ethanol is that the American farmer has not had a real marketplace to work in for years. The US government buys excess crops to give to hungry people in Africa, etc., even if they don't actually want more corn and wheat anyway, they get it from the US government. It's an effort to stabilize prices. And ethanol is just another way to do that. I don't think there's a corrupting force; I think there's a fundamental mistrust in the agriculture sector in this country to being subjected to market forces. I mean, we just simply don't believe in capitalism in agriculture. So I don't think that's corruption as much as they've unfortunately in some ways just fallen off the capitalist truck.

Now, there is corruption in the world obviously. You know, I was born in India and they clearly have documented cases of corruption there. But the thing with the clean energy economy is that it survives regardless of corruption. So, for instance, in India there are a lot of people who steal the subsidies for kerosene. The number that I've seen is 38.6% is that gets stolen from the kerosene subsidy pool. So only 60 some percent goes to the people. If you were to give people solar lanterns instead of kerosene, the same amount of corruption could occur but you'd still be pursuing a low-carbon future. So the same people could benefit from corruption in the low-carbon future that benefited from the high-carbon present. Corruption isn't the bottleneck. I mean, yes, we all believe that corruption should go away, but corruption isn't the bottleneck.

CS: That's interesting. Well, speaking of this, you know that there are people who don't believe that carbon and global warming cause one another.

JS: Well, this is the reason why I don't believe that climate change is actually an environmental issue—except maybe for tropical rain forests and deforestation. Black soot, for instance, is a pollutant, but carbon dioxide is not a pollutant; it never has been. So the environmentalists don't know how to talk about this issue without making something toxic. So they're saying carbon dioxide is toxic.

CS: I suppose it depends on how you define "toxic."

JS: But it's the wrong frame of mind. Whether you're right academically or not is beside the point. You're not going to convince people that carbon dioxide that comes from their own lungs is toxic. And so the answer has to be that this is all interconnected. No one denies that we lead a high-carbon lifestyle; what they deny is whether or not it actually causes climate change—whether it results in all sorts of unsustainable practices that have nothing to do with climate change. So for instance, we are using up fresh water at alarming rates to pursue our high-carbon lifestyle from the tar sands to coal plants. We are putting ourselves in a situation—like we did hundreds of years ago with salt—in our dependence on foreign sources of energy. Whether you're importing coal like India is or importing oil like the US, both are fairly scary situations when you have to depend on other countries for your economic growth. Resource constraints are real. Renewable energy technologies are ubiquitous, whether it is wind energy in some parts of the world and solar energy in other parts. The argument we're making at the Carbon War Room is that for 50% of carbon emissions technology and policy are not the problem but market implementation is really the problem—and maybe some micro-policy issues. This transcends whether or not you believe that climate change is real.

CS: I spoke not too long ago with James Woolsey, by the way, who may be the most visible spokesman for that concept that it doesn't matter which of these things you think is going to kill us first—there are 8 or 10 different reasons to do this.

JS: Right, but the challenge is not stating that there are 8 or 10 reasons to do this. I think Tom Friedman states this in his book—
Al Gore say this in his book—the challenge is trying to pinpoint exactly what the bottlenecks are, what the market dislocations are, and then finding how you would actually approach using your finite resources to make changes in that market to solve those
problems to unlock the potential.

CS: Well said. Now, it sounds to me as though this is "technology agnostic." In other words, this supports renewable energy and electric transportation generally. Is that correct?

JS: Yes. We are looking across 25 areas within seven theatres transportation, electricity, buildings, industrial gases in cement, steel production, deforestation and agriculture, emerging economies, and climate intervention. We are looking for places where the technology isn't really a problem and that policy is supportive and yet you still are not seeing solutions scale up. We do not expect to intervene in all 25 areas; we expect to intervene where we see the formula that fits our strengths.

CS: I happen to be a fan of solar thermal. A solar thermal farm in the US southwestern desert 100 miles on a square would provide more than enough power for the entire continent of North America.

JS: Yes, except it needs transmission, which may never get built. It's a human rights issue. The reason why China can build a transmission line is because they forced 15,000 families to move one half a mile from their home so that they can build it. In the same way that the United States government built a railroad across the country and did the same thing as shown in our Western movies.

CS: But we do have the concept of eminent domain that we've lived with since the inception of the Union 235 years ago.

JS: Eminent domain is weaker today than it was 100 years ago. I mean, people like Ted Turner, who is the largest private landowner in the USA is using their influence to slow things down for years.

CS: Right. And the Supreme Court just decided last week that the federal government had no jurisdiction over state and local government on cases like this thing.

JS: The thing is that 1880-1907 was its most innovative time period. When Nikola Tesla and Edison were duking it out between DC power and AC power and all this other stuff, Tesla invented a power system where you can move power through the air. We have not had innovation in the electric utility business since 1907 in any major way. So now, because of our inability to build transmission, we are allowing the space for innovation to occur.

Instead of communist style central command and control procedures using transmission lines to go from a central 10,000 square mile solar farms in the middle of the desert to every place in the country, we are going to see what the market will innovate in a low-carbon way to empower consumers to share power with their neighbors. I mean, I could generate excess power with a 100 kilowatt fuel cell in my basement, use the hot water from cogeneration from that for my family and power my entire neighborhood. I mean, that's innovation.

And the fact that we can't build transmission is going to lead to that innovation occurring faster. And when people say "Jigar, we're going to have rolling blackouts", I say bring them on, because that's what causes innovation.

CS: This is a fascinating perspective. So your overall statement is that capitalism is truly the answer. In other words, if you get out of the way of people's incentives to innovate when they need to and to make a profit where they can, good things will happen.

JS: Exactly. Yet we have to be cognizant of the fact that this has to be done in a way that is regulated by the government such that we keep the rights of the poor and the safety of the many in mind while we pursue this. But subject to that, I think capitalism is by far the best way to innovate our way to a low carbon future.

CS: Fantastic. As always, nice to talk with you Jigar. Thanks so much.

JS: All right my friend, it was great to talk with you.

Note: Have you even hung up the phone and later realized that there was something important that you forgot to say or ask? Here, I wish I had asked Jigar about political philosophy as it applies here.

In particular, it seems to me that there is a clear distinction between entrepreneurship and capitalism. This is illustrated by the inevitable conflict between the pioneering and independent spirit of entrepreneur, and the entrenched monopolies that always develop as the end-point result of capitalism, that make every effort to pocket, crush or sabotage the most potentially competitive entrepreneurs by any means at their disposal—and generally triumph.

One has only to point to the historical examples Nikola Tesla's DC power, Henry Ford's original ethanol engines, the Tucker automobile, the Red Car trolleys of Los Angeles, and the metal hydride battery, to show the harm that dominant power structures will do to creative thinking, but will dine richly on slave labor, child labor, concentration camp labor, prison labor and sweatshops. Meanwhile, we have to live with their rampant waste and pollution of ecological resources, unsafe and inefficient vehicles, clinically worthless medicines (to the exclusion of curative research), and harmful food-like products that simultaneously fatten and starve.

I wonder what Jigar would have said? I'll have to remember to ask him next time we speak.

For more information on this contributor, please visit: http://2greenenergy.com/renewable-energy-facts-fantasies/.

USING THE LAW TO ENFORCE ENVIRONMENTAL RESPONSIBILITY

The Natural Resources Defense Council's purpose is to safeguard the Earth: its people, its plants and animals and the natural systems. NRDC is generally recognized as the nation's most effective environmental action group, combining the grassroots power of 1.3 million members and online activists with the courtroom clout and expertise of more than 350 lawyers, scientists and other professionals.

The organization is probably best-known for its strength as a litigator—using the court system to force public and private entities to conform to laws that are in place to protect the people and the planet we call home. Johanna Wald, who has been with the NRDC for 35 years, spoke with me about her work there and her observations on the quest for clean energy.

Craig Shields: Let me begin by asking you this. I think that people have a sense that big government and big energy have a corrupt relation, and that this results in renewables' being suppressed. We need a kind of "watchdog," and that's how the NRDC functions. Is that correct?

Johanna Wald: I strongly think your thesis was certainly correct in the Bush administration, but I think it's less true here. I think that this administration is in fact trying hard to promote renewables, though I don't know that I'd say there is a level playing field yet. And I'm sure that the renewables people wouldn't either.

CS: OK. Please explain the NRDC's overall role.

JW: There are many parts of NRDC. And certainly a watchdog function, as I think of it, has been traditionally and historically a big part of our activities. My work, in particular, has been in opposition to fossil fuel development on public lands. I've been at NRDC for more than 35 years now.

CS: Oh my, that's great.

JW: My whole career has been devoted to defending the federal public lands—those that are managed by the Parks Service, the Bureau of Land Management, the Forest Service, and so forth. While I am also trying to prevent harmful and irresponsible renewable development, I am at the same time affirmatively trying to promote well-sited development.

CS: Wonderful. Now, if you don't mind, I think this would be more illustrative if we had an example. Is there a case in which this was going the wrong way, where someone had done—or was attempting to do something that was obviously corrupt, and the NRDC got wind of it and litigated it successfully, to a good end?

JW: Actually all the years of the Bush administrations would be good examples of that. The Bush administration, in its zeal, and it really was I think quite zealous about this too, put much of the nation's publicly owned energy resources in the hands of oil energy companies. It was out there leasing, I almost want to say willy-nilly, federal lands to energy companies, regardless of the other non-energy resources, the wilderness qualities, the wildlife, the water resources, regardless of the impacts that this would have. And we have been involved in a number of very successful lawsuits that could be used as examples.

On the renewable side, it's still too early. The Bush administration, while they were pouring staff time and federal resources into the oil and gas-leasing program, was essentially ignoring the renewable energy program until the very end of the administration when they started thinking about some issues. And they created quite a horrendous situation, certainly in California, by their lack of attention to this issue. Specifically, they did a wind environmental impact statement that was designed to establish policies and guidance for the wind program, but for the most part, their wind program just said that anybody could file an application. They did put some places off-limits—but for the vast majority of the BLM (Bureau of Land Management) land, the Interior Department said that the wind industry could file an application and they would work through it on a project-by-project basis.

There was, in other words, no effort to direct or manage where these applications would be filed or to get them to the least conflicting areas. That's actually the way they ran the oil and gas program; they were clearly using it as a model.

When the Obama administration came into office there were more than a hundred applications for utility-scale solar projects in the state of California alone. And the environmental community thought they had to accept some applications in some very bad areas; they hadn't completed the processing on any of these because there were way too many of them, they had no guidance from Washington DC, and they had very little expertise. These are almost all brand-new technologies and at a scale that we've never seen before in this country, and they didn't have any human resources, because the Bureau had been told they had to "do more with less." At least with oil and gas they had more resources. So that's the situation that the Obama administration inherited.

And I have to say that compared to the Bushies, they have been much better. First of all, they understood that, with respect to solar, you have to manage it. You can't just say anybody can file anywhere and we'll do it on a case-by-case basis. You can't do that for two reasons. One is because we really do need to get some of these projects online so that we can do something about climate change and bringing down our

CO_2 emissions, but also because that isn't the way to get things online without controversy, which means delay.

And indeed the environmental community was rightfully concerned about some of these projects. The siting of them, the potential impacts—it's true these projects will help us address climate change but they are certainly going to have some significant environmental impacts.

So the Bureau has taken a great approach to their solar programmatic EIS and to the development of their program. They're looking for areas that are appropriate for solar development. They're specifically looking for areas that have low conflicts, and they're conscientiously avoiding places that they know from their own experience are the places that people will fight to protect and preserve, and they're looking for places in these low conflict areas that are near transmission lines and that already have roads. And then they'll try to prioritize development in those areas, which will allow them to focus their efforts. And, equally importantly, by focusing efforts and focusing development, I call this clustering development, in appropriate zones, that will help them allocate their resources. And they've gotten more resources to do this work from the Obama administration. This will minimize the proliferation of projects across the land.

So that's where they are now. This wasn't going to be a very easy task for them under the best of circumstances, and we don't have the best of circumstances...we don't have the best in several ways. The first is that, as I've said, there are already a lot of these projects out there. And what will happen to those projects after the EIS and the program is established is somewhat unknown, so they have to do something about the projects that are pending now.

Also, we have the stimulus bill, which held out the possibility of significant funds being made available to renewable project developers in solar, wind, and geothermal who can get their processing done by December 2010. This has created huge pressure on the Bureau of Land Management in the states where there are projects that potentially could meet this development—and there's no place that has more of those projects than California. In California the pressure is not only on the Bureau, but it's also on the California Energy Commission, which by state law has to permit solar-thermal projects which are greater than 50 megawatts.

So this is all background so you know what NRDC is doing about this situation. First of all we are working very hard. We work hard to get the Interior Department to adopt this approach to the solar program with zones and focusing and prioritizing. We work hard to get them even to think about doing something like that and we will be working through the programmatic EIS process to get them to actually adopt that kind of a program when it's finished. That's very important to us. We are working with other members of the environmental community to help them, to get them first to join with us in promoting these policies, but also to help them understand and appreciate the tradeoffs that are involved between renewable and non-renewable development.

Some of my colleagues here at NRDC have been promoting conservation and energy efficiency for decades. They're working on distributing generation and decoupling and other policies that are necessary for the utility industry to go from being based on fossil fuels to renewables.

CS: So you must be feeling pretty good about the way things are going, then?

JW: I'm cautiously optimistic. The pressure that the agencies are under is enormous. As I said, we're talking about brand-new technologies here. The regulators have no experience at all with these projects at the scale we're talking about here.

The watchdog function for NRDC in this context is really clear: it's to make sure that the best possible places are chosen. We need to ensure that the agencies, to the extent that they can, are cautious and respectful of the circumstances in which they are engaged here. On the one hand we do have to move ahead, on the other hand we have to do it in a way that will allow us to take advantage of the information that we obtain from any projects that are permitted.

CS: Yes, I'm with you all the way. How fascinating this is. Thanks very much for this.

JW: Yes, it is totally fascinating. For me it's totally fascinating because it's very proactive, it's a great challenge.

For more information on this contributor, please visit:
http://2greenenergy.com/renewable-energy-facts-fantasies/.

THE ROLE OF NON-GOVERNMENT ORGANIZATIONS—NGOs

For many decades, the World Resources Institute has assisted leaders in both the private and public sectors in making sound decisions with respect to practices that affect our environment. Here is my interview with WRI's Vice President for External Relations Robin Murphy, from the organization's headquarters in Washington, DC.

Craig Shields: Do you want to begin with a mission statement for the World Resources Institute—for people who may not know too much about you?

Robin Murphy: We work at the intersection of environment and human needs. WRI's mission is to provide analysis, research and recommendations, that will help advance sound environmental decisions. And those are decisions made by leaders in business, leaders in government, leaders in fellow NGO's—non-profits around the world, in academia and elsewhere.

When we talk about how we work at the intersection of people and the environment what we're talking about is that there are human systems on this Earth, we being the predominant species...and then there are natural systems. And while we've exerted a fair amount of control over some natural systems, we'll never exert control over all of them, though we are having an impact on them. And what we try to do here is to look at and figure out how we can have them work more in balance with one another so there is sustainability for people on this planet. And this means digging into the data and determining how these systems can be altered, calibrated, and rethought—so that we can deal with a burgeoning population, with climate change and so many things that are coming toward us. We help address all these issues in rational ways that are equitable and sustainable.

CS: That's wonderful. How are you funded?

RM: We are a non-profit organization, and we receive funding from four sources: individuals, foundations, governments, and corporations—and that proportion varies from year to year. A couple of years ago we had a much larger proportion of corporate grants. The pie has thankfully gotten bigger, but now it's more government grants. So it varies from year to year.

CS: OK. It sounds like an important mission. Let's discuss it in the context of renewable energy. Where do you see your institute's activities as they apply to moving to renewables?

RM: Our point of view on this is of course not to pick and choose. There are a lot of possibilities out there in terms of renewable energy and of course the wedges have defined a great deal of what's possible there. Our work is in helping with public policy that will encourage the best use of renewables as they're developed. Some have unintended consequences and so we have a role in pointing out with empirical evidence how various approaches or solutions stack up in terms of their environmental impact—in the short- and long-term.

CS: To the end of recommending the right course to policy makers?

RM: Yes.

CS: I see. One of the reasons I ask is that I've spoken with people at the Audubon Society and the Wilderness Society and so forth—and that's pretty much uniformly what they say. Yes, we're in favor of renewables but no, we're not in favor of blindly tearing up wilderness where you could have done the same thing 100 miles away where there's already a human footprint. So that's the type of thing you're talking about I presume.

RM: That's precisely what we're talking about. What may appear to be the perfect solution may not have been thought through and there may be unintended consequences or things that really haven't been taken into account.

CS: How would you say that the trajectory for renewables should be different in developing nations than it is in the developed world?

RM: Very often the people who rely the most on clean ecosystems are the ones who have the least access to decision-making about them. WRI is concerned about the unintended consequences of development. Large-scale development decisions are being made, building dams and other things like that, without sufficient thought being given to the cascading consequences: to the environment, to people, to public health, to the quality of water and the quality of ecosystems. There are so many people who are affected by these mega-decisions that are made and they can be very blunt instruments at times. They can be as well intentioned as possible but they may not have taken into account the vicissitudes of governance, for instance—of corruption, of geography, of economics— there is a whole host of things in which we have expertise where we feel we provide a valuable contribution.

CS: Let me ask you about Copenhagen—which, by the time this book is published, will be old news—albeit probably not very positive old news. But the reason I bring it up is that ironically, the people in the developing countries are going to be most affected by a cause that they had nothing to do with. In other words, the hardest hit, the most easily displaced, the most easily flooded and beset by famine are the people who had the least to do with this problem in the first place. Do you think that's true, and if so, what does WRI think about that?

RM: We do think it's true. It's painfully obvious; the people who have not contributed to global warming, who've not had industrialization will suffer from the direct consequences of it. The floods in Bangladesh, the people who live on the Maldives Islands and other places are not putting out tons and tons of carbon. There are a lot of people aspiring to have a better life, however that's defined, with at least security and comfort and shelter.

And so yes, it's a huge concern. Many people call for "environmental justice." The statistics don't lie about it. We've measured per capita emissions, and when you know the United States is at a certain level of 20 or 21 and the average Chinese is 4 or something, there are huge inequities. You do begin to look at this as a number of people in the religious community and elsewhere—it's not just scientists and policy makers, it's people who have a real concern about basic human equity.

It also presents huge complications; you have a lot of things that are baked in, that are very difficult to reverse after decades and centuries of history, out of habit and cultural norms, and so these are some of the huge barriers—changing frames of thinking. One of them is the value of ecosystems services—for a long time we had the luxury of being extractive and we could kind of take whatever we wanted. Well now, how do you go about dealing with limits—and do so equitably? And how do you do it so that economies and individual businesses can thrive. I mean this is not an anti-market or anti-capitalism concept at the root of things. But how can you introduce those kinds of better sustainability and a different way of thinking about ecosystems and their health so that people and economies can thrive?

CS: That's an excellent question. Could you tell me more about ecosystem services?

RM: They are just simply air and water and trees and the respiration of oxygen and the whole carbon cycle. There was a thing called the Millennium Assessment, done under the UN umbrella, and they took 24 major ecosystems and assessed them—literally like a traffic light, red, yellow, and green…most of them unfortunately came out red, meaning degraded. Yellow meant under stress. And green was healthy.

And we've picked up the mantle of this Millennium Assessment and have used that as a kind of template to put together a number of reports that we've disseminated to businesses, to municipalities, to mayors and to others to make them more aware of the value of ecosystems services.

Coca-Cola learned this the hard way—it takes seven gallons of water to make a gallon of Coke. And so they have to be located in places where there is an enormous amount of water and there's huge discharge. And a number of years ago, they went cross-wise with the Indian government and got in a lot of trouble over there, but they now consider themselves a water company. They're a soft drink company but they have to look at water and water supplies and water sources, otherwise they can't do business and they can't produce the kind of product that takes this into account in terms of the health and vitality of ecosystems for everyone's wellbeing. You make a pressure point here or you destroy something here, it can have a cascading effect that can cost you more in the end, down the road, unless you've thought it through beforehand.

Here's another example. The city of New York a number of years ago was about to spend literally billions of dollars to build a huge water treatment plant outside of the metropolitan area. And somebody pointed out that actually if they bought up huge amounts of land on the upper Hudson, where there's still forest and simply preserved the forest area up there, that the natural system of keeping the water pristine and not dumping into it or allowing development along the shores would allow them to have a clean water system coming downstream into the city. So they made the decision instead to pay land-owners to hang onto it and to keep it in that natural condition because it's already a natural conditioner.

CS: Great stuff. Where do you focus most intensely to bring about a fair and positive change?

RM: Well, we can't make it all happen. I have to say that we do take on the issues that we feel are the largest and contribute what we can to them. And that was a very deliberate decision from the very beginning. In the early years of this place, in the mid-1980s, climate change and ecosystems services were the two main things we selected, and we're still focused there. By no means do we feel we have the lock on everything, but we try to find the areas where we can get the most purchase—the most traction. Putting out academic papers is nice but that can get you nowhere. How can we find the levers that will make change? Who are the champions who'll make change, for instance, the leaders of business who do have this kind of foresight? And they don't see it from an altruistic point of view; they see it from a market point of view, a share-holder point of view.

CS: Maybe you can give me an example of that. Are there cases where, through your work, you've gone to one or more business leaders and said "Look, here's something where we can all win?"

RM: I'll give you two examples, Craig. One started around 1998 I believe. And it's the greenhouse gas account protocol. Back then, some businesses and other leaders were beginning to see that they were going to begin to work in a carbon-constrained world economy. And so how do you measure emissions? If you can't measure it, you can't make any adjustments. And so, WRI with the World Business Council for Sustainable Development, which is based in Switzerland—a fabulous group—and Arthur Andersen, which is what it was at the time, got together and got our experts around the table and said, "How do you set up a 'FASB'—an accounting system?" We want something that can be generally accepted and credible and standardized so that a ton in India is the same as a ton in the US. A ton in an auto plant in Japan is the same as an auto plant in Italy. So you have standards of how you go about measuring greenhouse gas emissions. And over time, this has become the norm; it's the standard now. It's generally accepted and used throughout the world.

And our people here are now drilling way into the weeds, because that's what it's come to in terms of how you begin to measure things that are esoteric or arcane—or frankly that are just new—new processes that people haven't thought of. For instance, throughout your supply chain, how can you account for greenhouse gases among your suppliers? How do you put in land use? How do you account for deforesting a place and determining how much you're upsetting or allowing more emissions to go into the atmosphere when you've just cleared several thousand acres in Indonesia?

The greenhouse gas is a technical contribution that I feel is also actionable and practical. That is the sweet spot of what WRI tries to do—to come up with something big, to work with partners on it, because we can't solve it all ourselves, and then start to implement it and get it out there so that it takes root.

The second example is the US Climate Action Partnership, USCAP—a group of 35 or so organizations. Seven of them are environmental groups and the rest are Fortune 500 companies throughout a number of sectors. In auto we have Ford and GM, in energy we've got PG&E, Duke, and Florida Power & Light; in chemicals we have Dow and DuPont; in extractive industries we have Rio Tinto. This whole thing came about when my boss Jonathan Lash and Jeff Immelt at GE had become friends. John helped them think through their "ecomagination" thing years ago. They really felt that getting federal law written—public policy at the federal level—would benefit from more than just hearing about the science from environmentalists like us. They asked why we don't work together and make recommendations that pull together the scientific aspects of this with the practical implications for business. How is it going to have an affect on an electric power company or automobile manufacturing or a chemical company like Dow? What will it do in terms of supply chains: shareholders, employees—the whole range of things?

They've been working now for about four years in putting together this group, and they have produced two very public blueprints that have been presented to Congress. This is all online; I can send you the links.

You'll see lots of working papers and lots of meetings—all this, frankly transparent. I wonder if this has ever taken place before in the history of the United States, where you had this kind of an incubator working alongside Congress. It's completely volunteer; nobody asked them to do it, except they just all decided they'd get together and do it.

CS: Well at the risk of appearing rude, I would point out that corporations get together or work singularly to influence congress all the time. It's called lobbying. What's the difference here?

RM: Our answer is that it's called capitalism. We've been accused, publicly by senators, of being "pigs in a trough"—and I can see how some people will look at it that way. I guess the point of view that we have, is that it's not about individual companies, there are so many sectors—the oil companies are in here too—there's BP and Shell and others. There are so many sectors that represent trillions of dollars in equity, hundreds of thousands of employees—it's not some small little crowd.

To critics who say "You're just in it for yourselves," frankly there are plenty of others who could join if they want to or be part of this kind of a process; we just happened to kick-start it. And nor do we feel that we have all the answers. There's a lot of argument around cap and trade for instance. We just simply have the point of view that changing markets is the way to go in this, rather than a carbon tax, which is so prone to political manipulation.

When you create new markets, they're pretty hard to change because that's self-interest and they have a self-regulating effect. From an internationally competitive point of view as well, we have to get moving on this. These CEOs at USCAP constantly say this; they say this publicly—that they are holding up all manner of investment because they don't know which way to go; they want to know what the playing field is—what the boundaries are—what the rules are. The federal government sets the rules and then there's everyone getting into the fray. I'm straying a bit from your question but mind you, this is pure capitalism.

CS: Now I understand more about what you're saying. Let me ask you about subsidies. Some people say, about corn ethanol for instance, that if we hadn't done this thing at the behest of a few extremely powerful senators in the corn-belt, we'd all be a lot better off. Where does WRI stand on things like this, and promoting truth and transparency?

RM: We really try to approach it from a scientific standpoint. We have an "institutions and government" program and they have a thing called the "Access Initiative" that they run throughout the world. And the main focus is working with a lot of in-country civil society organizations and other NGO's on how can you get more transparency in government. How do you get more participation in decisions that directly affect people—not giving away timber concessions, not trashing villages, etc. And so from that standpoint, it's taking it from the scientific analysis and assessment point of view and then asking how we can provide the kind of information and the kind of levers that will help make change.

When it's things like ethanol, these things are highly political and there are often special interests involved. That's nothing new of course. What we do is try to connect the dots and to say this was an investment that was made, here's an array of other things which would have had less deleterious consequences and here's an investment that would have much greater promise or fewer environmentally damaging consequences. So it's always about analysis and verifiable technical background research that we can do and bring to the table.

CS: Okay, so what you're saying is that, though you're not directly involved in the political process, you try to make this thing so obvious that, if corruption will occur, it will be forced to do so under the glaring light of scientific evidence. You can't keep it from happening, but you can make it pretty damned obvious.

RM: Yes, exactly. In fact we spent a lot of time on Capitol Hill. We have a staff member who does that. We take our stuff to anybody who is interested in the subject. Because again we feel like this kind of information analysis and synthesis is power. And if you're going to make an argument, you really can base it on credible evidence.

We don't have a membership like a lot of other organizations do—and sometimes I wish we did. It gives you a lot of power and a lot of clout because you have a constituency, but we don't have a constituency. Zero members. And so it has to rely on the credibility of our work, which is peer-reviewed, outside reviewed, etc.—before it's ever gotten ready for prime time. That's the contribution that we can make and the value that we have.

CS: Well thank you, Robin. This has been everything I hoped it would be.

RM: Thank you. I'm glad.

For more information on this contributor, please visit:
http://2greenenergy.com/renewable-energy-facts-fantasies/.

TOUGH REALITIES

THE TECHNOLOGIES

BIOFUELS

I hope the preceding section has presented a reasonable case for the imperative to move in the direction of renewables. Having said that, I'm quick to point out that most of these discussions are not at all black and white. In particular:

A professor at MIT is soon to release a paper pointing out that CO_2 is a significantly weaker greenhouse gas than was previously believed, and that increased concentrations in our atmosphere will be likely to cause only a fairly small increase in the Earth's surface temperature. But what will be the effect of forces that could serve to accelerate global warming, like the melting of the arctic permafrost and the resulting release of the methane under it? And what about the reduced concentration of aeresols?

Let's be honest. No one knows. To say, "The debate is over" seems very foolish and wrong to me. What, you don't want to be confused with facts? I think rational, unbiased people have to acknowledge that there **is** a debate on the severity of the problem here—a debate that is composed, unfortunately, of 10 parts bombast and power politics—to one part honest science.

The oceans are a huge "sink" for CO_2, and serve to mitigate the increase in the rise of CO_2 levels in the atmosphere, even as we continue to destroy the rain forests and burn hydrocarbons. Of course, this comes at the expense, most believe, of increasing the acidity of the oceans, damaging the development of shell-fish and the entire ocean ecosystem. Exactly how much? Again, no one knows.

Other recent developments include the idea that some—or even most—of the oil on the planet is not fossil fuel, but was made abiotically when the Earth was formed. What will be the result in terms of the availability of oil and natural gas? We recall that Matt Simmons said, "It's unclear if we're running out of oil; but we're certainly running out of cheap oil." If there are huge natural reservoirs of such hydrocarbons very deep under the Earth's surface, how does it affect the discussion? Once again, it's anyone's guess.

We need to deal with the fact that there will be ongoing controversy about the effects of extracting, refining, and consuming carbon-based fuels. But here's a point to ponder: Our rates of childhood cancer are going through the roof. Does anyone have a suggestion for the cause, apart from the fact that our way of life here has become so completely unsustainable?

Additionally, in terms of the geopolitical and macroeconomic elements, consider these hard facts: As of December 2002—when the military pre-planning and pre-placement of ammunition and troops in preparation for the Iraq War had begun in earnest—there were about 972 billion barrels of total oil reserves in the top sixteen oil nations. Of this total, 69.98% was inside a turbulent Middle East, a mere 12.64% was in the Americas (South and North combined), and only a piddling 2.33% was here in the US. Over 80% of the world's oil reserves are within the territory of the top sixteen countries.

Here's a 2002-03 breakdown of those sixteen countries' oil reserves from the US DOE:

Rank	Country	Billions of Barrels Reserve	% of the Top 15
1	Saudi Arabia	261.8	26.93%
2	Iraq	112.5	11.57%
3	Kuwait	98.9	10.17%
4	United Arab Emirates	97.8	10.06%
5	Iran	89.7	9.23%
6	Venezuela	77.8	8.00%
7	Russia	60.0	6.17%
8	Nigeria	32.0	3.29%
9	Libya	30.0	3.09%
10	China	23.7	2.44%
11	USA	22.7	2.33%
12	Qatar	19.6	2.02%
13	Algeria	13.0	1.34%
14	Mexico	12.6	1.30%
15	Norway	10.3	1.06%
16	Brazil	9.8	1.01%

Note: Canada has 180 billion barrels of reserves, but 174 billion are locked in tar sands, and extracting it is highly inefficient and expensive.

With approximately 4.6% of global population, we here in the US currently consume just short of 20 million barrels of oil a day—that's imports and domestic production combined. We gulp it in for our agriculture, transportation, industry, electricity generation and home heating, in the form of various fuels, petrochemical fertilizers and pesticides. We also use it as raw material for a wide range of plastics, solvents and medicines. At that furious rate of consumption, if we're ever bereft of foreign oil, our domestic 2.33% signifies an interval of less than three before we consume all US reserves.

The two most populous nations in the world, India and China, each have over a billion people. They're chasing our consumptive lifestyle and our infrastructure at a breakneck speed, and oil simply cannot be sucked out of the ground fast enough to slake that thirst.

—

We now come to a set of interviews on a few of the popular technologies that represent the constellation of renewables at this point in our evolution. Again, I was lucky enough to have found subject-matter experts who, I believe, did a great job of articulating the basics issues of their chosen area.

We begin with Dr. Gregory Mitchell, Research Biologist and Senior Lecturer, at the Scripps Institution of Oceanography—one of the oldest, largest, and most important centers for ocean and earth science research, education, and public service in the world. Dr. Mitchell was good enough to speak with me about the use of algae as a biofuel.

Craig Shields: Thanks so much for helping with this chapter on biofuels, Dr. Mitchell. From a high level perspective, we're converting the chemical energy that was ultimately derived from the sun in the form of high-energy carbon bonds and we're breaking them down, burning them for instance in biodiesel, and converting that to kinetic energy. Is that essentially right?

Greg Mitchell: Yes, that's right. Fundamentally, the photosynthetic process reduces inorganic compounds like carbon dioxide, water, nitrogen, and phosphorous, and builds these biochemicals. Initially sugar, and then the sugar's burned to build all sorts of other things, and nutrients are brought in and you build membranes with phosphorous and you build proteins with nitrogen and so forth. So yes, it is all ultimately derived out of the sunlight. Then, if you dry out any organic matter, you can burn it; you take plants, cow dung, etc. So that's called oxidation, of course.

It's a cycle. You take oxygen and water and inorganic carbon dioxide and nutrients and you use solar energy to push them up the energy chain. Once they're up there, they have potential energy and then if you oxidize them, which is effectively burning them, then you bring them down. Mammals' warmth comes from the slow burn of these molecules that releases heat, and then we insulate ourselves and hold onto that heat to be warm.

So yes, you've pretty much got it right in that regard. What people need to understand is that petroleum, coal, and natural gas are the same category of thing. They are biofuels. They are carbon, organic carbon, fixed in some ancient time usually between ten million and a hundred million years ago, fixed by the same photosynthetic process that we're talking about now for "biofuels." If you really think about it logically, they're all biofuels.

So there needs to be a clear distinction between biofuels derived out of the contemporary atmosphere and then reburned into contemporary atmosphere and biofuels that derive CO_2 out of an ancient atmosphere and then stored it away in a geological reserve and then it's exploited and burned into the contemporary atmosphere. It's just that now we have the carbon cycle disrupted by our burning rapidly a huge reservoir that was put down millions, tens of millions, a hundred million years ago. So we're burning it at a pace far faster than it was ever laid down geologically. That's what's causing the imbalance in the atmosphere.

CS: OK, that makes sense.

GM: So, now the kind of question is, "Well, can we cut that time cycle short? Can we just take the solar energy now and use it?" Of course we can. We can and we do. Of course, people have chopped down trees and burn them throughout the millennia, and that's one of the big reasons the Earth's becoming deforested. So not only can we do it, but we have to ask what are the environmental impacts of what we do now, and what are the unintended consequences. If we just start cutting down all of our forests to replace petroleum, now we deforest the land and we've got other problems besides the fact that we're buying petroleum from countries that don't like us very much.

It's very complicated. So where do you start? I think the first thing you do is you ask questions. What do we humans need to survive? So, everyone's saying "Well, we need energy," and that's true. It's very clear that industrial capacity, standard of living, educational levels, are all proportional to the rate at which any nation consumes energy per capita. A nation like Japan or Europe might be much better than a nation like the United States in terms of having a similar achievement level economically and socio-economically, and yet utilizing less energy, but California's on a par with them. We've got thirty something million people in California, and we're on a par with all those other nations. But there's other places in our nation that aren't on the same par, like Texas, that are farther behind per capita. This relates to policy. So we do need energy, and energy drives our economy, and we need food.

The second one I'd like to stress that people don't pay attention to that much is that 80% of our agriculture is to feed our animals and 80% of water use on Earth is for agriculture. So something like 60% of our water use is to feed animals. And if you start looking at energy, which is, I think, the overall premise of your book, right?

CS: Exactly.

GM: You need to look very closely at the consumption of energy for agriculture and for pumping water. In California, about 30% of our electrical use is to pump water, and more than half of that is for agricultural and irrigation and distribution purposes. A big part of that energy consumption is to pump water over hill and dale to get to the sewer plant to throw it away into the ocean. So there's all sorts of ways we could go upstream and not pump over hill and dale and process our water back to potable water reuse, or certainly irrigation water, and pump less and save water. If you save water on the front end, you're using less energy on the back end.

My point is that the biological cycle and the water cycle and the nutrient cycle and the CO_2 cycle are all the same thing. They're all driven up by sunshine, and then it's all degraded back, mostly by bacteria, to inorganic nutrients. That cycle goes around and around—and so when you start talking about energy, or bio-energy, you have to talk about water and nutrients. And you have to talk about CO_2. They're all linked.

I think you could go to an ecology textbook 101 and you can see that water, inorganic nutrients, and sunlight make the organic molecules we talked about, and that respiration brings them back down. The fact that it's a cycle suggests that we can proceed into the future to try to tighten that cycle up as efficiently as we can, locally.

So, when you get back to biofuels themselves, what's the question? What do we really need? Well, we need energy. Biofuels are in everybody's wedge of how we're going to replace liquid petroleum. If you look at the requirement for liquid fuels into the future, every oil company and every nation in the UN has a biofuels wedge growing. At last year's annual meeting at the American Association for the Advancements of Science, AAAS, the talk about biofuels was depressing—if you are going to fill that wedge by 2030 or 2050 like is in everyone's chart, you're going to destroy the Earth. You consume the water, the rainforests are going to go, you just can't do it. There's not enough land to do it with traditional agriculture.

But your question to me is about biofuels. We clearly need to produce food with sunlight through photosynthesis to feed our animals or else we should all become vegetarian. And we could all become vegetarian, that might be useful, but I'm not sure that the two-thirds of the planet that's not vegan wants to become vegan. Animal protein is important for most humans.

So now you get to the question of efficiency. Here's a third main point I'd like to make. The solutions can only be achieved if we create wealth in the economy. In other words, you can't just tax the existing wealth base and solve the huge problems because the costs of the solutions are so huge. They have to pay for themselves.

When you ask "What can pay for itself?," that sounds difficult, but fundamentally you look for efficiency. So that takes you to algae, and also cyanobacteria, which are actually a form of bacteria that are photosynthetic, so I'll refer to these collectively as algae—photosynthetic organisms—mostly the microscopic ones because they're super-efficient at growing—at taking up nutrients and CO_2 and creating biomass with sunlight. The microscopic ones are more efficient, faster-growing than the larger ones. You go small. This class of organisms, photosynthetic micro-organisms, can produce biomass 10 - 50 times more efficiently than any terrestrial plant and they can produce protein on that scale, 10—50 times more, so in other words if you have a soybean field, you can produce about 10 times more protein per acre per year growing algae or cyanobacteria and about 50 times more biofuel molecules. So right now, soybeans are mostly grown to produce pig food and chicken food—you know they press the soybeans and the soy powder is pretty high in protein and it feeds our animals. And the residual oil is available for bio-diesel. So, if you have something that can achieve 10—50 times these yields, any industrial economist looking at creating wealth, if you find something that's 10 or 20% more efficient, they chase it with billions of dollars in investment to beat the competition.

Here we have an agricultural possibility that's 10—50 times more efficient. And not only that, Craig, but these systems can be laid out on non-arable land, so you don't have to worry about topsoil; you can grow them with saline aquifers or seawater, and so you can perhaps minimize the land impact and otherwise food for fuel debate because you set it up on land that's not otherwise suitable. And you also eliminate that debate by using seawater or saline aquifer water that you can't irrigate with or drink. So these are huge advantages.

CS: Wow—no wonder people are so excited by all this.

GM: Right. And for this reason, society needs to invest. Now can we make a profit on it tomorrow? No. But if you're going to go down this road, if you really do explore the cost of energy for agriculture, what you're going to realize is if we make algae, our solution for pig food and hunger then actually, we're going to save a huge amount of energy and water. To me, that, and this is my last, most important point, I think that is the most significant driver, and not fuel. It's animal feed, which if you look at the economics of algae, it turns out that the best you're going do in terms of biofuel molecules is maybe 30 or 40% of the original biomass.

It could be into a diesel if it's a vegetable oil, which most people are chasing with algae. It turns out algae can also make a lot of carbohydrate—and you can go down the biobutenol path or ethanol path or other paths that would derive out of carbohydrate, so you don't have to be exclusively into the vegetable oil diesel kind of a product. You could also be into the biobutenol or ethanol types of pathways as well, with different kinds...now it depends on which algae you grow.

So, that might be 30 or 40%, but what's the other half or so? It's protein! And that protein is what we need to feed to our animals in addition to some Bermuda grass or whatever the farmers grow.

So the point is that we might be able to offset a vast amount of agriculture with much less area, and that gets back to your question about how much area do you want for solar panels to get all the electricity. The same diagrams exist if you look at, let's say, replacing all of the US diesel by corn. It takes up the entire United States. We are replacing half the US diesel—and this is the line I got from Chevron— half the US diesel from corn would take all the United States. Half the US diesel on soybeans, which produces more oil, would be like a third of the United States. And half the US diesel on algae would be this little tiny snippet in Arizona somewhere.

CS: Wow. This as well came from Chevron?

GM: Right. This diagram is from them. You know, they have it, Exxon has the same diagram, all the big oil companies are looking at this.

It's economics I think, Craig. An investor's going to invest right now because he thinks he can make a profit in 3 to 5 years. And we're not at the point where the algae production systems are efficient enough, that we know how to do it well enough, to actually make a profit right away. Although some companies say they are there, my personal feeling is that if they are there, then they should be making billions of dollars a year. They're not there yet. But some of them may be close. From an engineering production scale point of view, it's clear we have not done it yet. But if you think it's impossible, it's not. Going to the moon looked impossible but we did.

There's no reason why we can't; there's no reason humans can't do it. If it's a scientific breakthrough required, there is possibly a major impediment, but I don't see the scientific challenge. We know the growth-rates, the yields; we know these things. There are interdisciplinary problems of all sorts, but I think it's a matter of putting it together and engineering it.

CS: What would you say are some of the engineering challenges?

GM: There really is a lot of work aimed at making the right decisions—at least a dozen. It's everything from strains that produce the bio-molecules of interest and can be sustained in a high density mass culture without being invaded by something else that takes over and not getting attacked by bacteria and viruses and suddenly just crashing on you in ways you don't understand. These things relate to the organism of choice. We need to find elite organisms that work.

In other words, not everything works for terrestrial agriculture. In fact, surprisingly few plants on Earth have ever been domesticated to make a huge difference. You know, we have corn, sorghum, soybeans, whatever, there's a huge number of plants on Earth but not that many of them have really been domesticated. So we need to find these elite organisms that are going to perform super-well, and then we need to have systems that are very low cost that we grow them in. We have to bring the production cost down by a lot. So that's an engineering challenge.

Then the next challenge is—though it's probably not as significant—separating the algae from the water. There is a lot of energy associated with moving water around, so we really need to find very efficient ways to pull the algae particles out of the water without spending too much energy; that's sort of the harvest step.

Once they are harvested in high concentration, you've got to bust the cells up; you've got to extract the molecules of choice, you've got to separate things, and process them—refine them. Once you concentrate the material out of the water, then it becomes basically the food processing and petroleum refining industry and the beer industry, they know how to deal with these kinds of things at the downstream. So I actually don't think those will be that big of a problem. I think it's the front end, which is the elite organisms, bringing the production systems down substantially, and finding a very low energy way to harvest.

CS: Didn't I hear something about an ostensible breakthrough in algae recently, feeding plants macro-nutrients instead of just exposing them to sunlight directly?

GM: Yes. There are companies like Solazyme that are talking about growing algae in fermenters with organic molecules. And there is another company named Martek that already has a large business fermenting algae to make polyunsaturated fatty acids that go into baby food, for example; they're putting the stuff in there because it's good for brain formation for babies. They have a $300 million market selling high-value products. Pretty much almost anyone I know in this business believes that fermentation of algae to high-value products is certainly viable for an industry like Martek, maybe Solazyme, but I don't see how you can go to scale for fuel for that.

CS: That makes great deal of sense. So what needs to happen to move this industry along?

GM: This is where the public has to come in and back the research to get us there.

How did Brazil become an ethanol economy? It was a major

commitment by the public. The government made a decision to do it, and they did it. It took them 20 years. So is our government or any other government going to make that decision about algae? I would strongly recommend it.

CS: Fascinating. It goes back to the crux of this book—looking at the practicalities—and the impediments. In other words, there are technological, there are financial, and there are political impediments in each of these areas. I would think the people in traditional energy with the "drill, baby, drill" mentality don't like this.

GM: Well, "drill, baby, drill" is a misperception that there's somehow this opposing dynamic. The charts from the oil companies pretty much accept that they are going to be able to produce about 100 million barrels per day. Even if they "drill, baby, drill," they can't produce more than that, but demand will go up. We can't fulfill the demand that society is going to put on the overall economic system, and there's also these major climate challenges, so we have to do two things. One, is create alternatives that will fill in the demand because there are six billion humans on Earth and only one and a half billion actually live relatively prosperously, consuming lots of energy. The other four and a half billion have growing aspirations. This demand won't go away.

No matter how much you "drill, baby, drill," you're not going to exceed a certain maximum and then you still are going to have demand above that. So I think what all these oil companies are looking at is "We're going to have to fill this wedge in," and biofuels are the wedge. See what I'm saying?

CS: Absolutely. But how do we get from here to there?

GM: The impediments are similar to those that are in front of various emerging technologies. Let's just take the broad-based desire to develop a biofuels industry in the United States, in part for security; it doesn't have anything to do with the environment—it really has to do with security of our liquid energy supply. And so there are tens of billions of dollars—somebody told me it's $20 billion a year of subsidy for corn ethanol. Okay, well everyone knows the corn ethanol is bare minimum, breakeven, in terms of CO_2 and greenhouse gases—maybe even negative, in other words you produce more greenhouse gases than you offset. And everyone knows that you can't do it cheaply enough to equal petroleum. You've got to have a subsidy.

If we invested the amount of the corn ethanol subsidy in algae, we would create sufficient scale to drive the cost down by about a factor of 5 to 10 from where it is now, which would be huge, because then investors would start chasing it. But there is a lack of knowledge of the potential and a lack of confidence that public policy investment, which means money, would truly overcome the hurdle. So one of the impediments is a perception among the politicians, which means a perception in the public. The public doesn't understand it.

That's the role you play, Craig. I really think this is one of the most significant impediments: the lack of knowledge of the potential, and a lack of commitment by the public to endorse this as a substantial option that should be done.

But here's the root of the issue. Suppose we create public understanding. But you know what? There is no economic constituency for algae. There is for corn and soybeans. So now you go into policy. No matter what the public believes, the policy-makers are listening to where the jobs are, right?

There are no jobs in algae. We don't have industry. So that becomes a huge policy impediment.

CS: How is that, though? If you're using non-arable land, and you have all this other good stuff going for you. Isn't there a legitimate trajectory for jobs?

GM: There are a lot of jobs to be created in a 10 year time-frame, but a politician's lifetime is not 10 years. It's two or three years. So they can satisfy a huge number of constituents right now that depend on soybeans or biodiesel infrastructure for corn or tractors, all these things that relate to traditional agriculture. So there is not a large industry that's built around algae, and in the first five years it's going to be a relatively small numbers of jobs.

CS: I see. This "political cycle" seems to be a recurring theme in the interviews I've conducted.

GM: I don't doubt it. I think that the real impediments are going to remain forever if the public and the policy-makers don't get on board. So I actually think the biggest impediments are the public and the policymakers and then, if we have the funding for 10 years, and the patience, a highly mature industry will follow.

CS: Great. Let me ask you about the allocation of funds under the stimulus package. Has algae received anything at all? Obviously this goes back to the subject you were talking about a couple minutes ago. But what percentage of overall kind of renewable energy dollars have you received?

GM: I don't know the percentages, but I do know there is substantial money that is now moving into the algae space. And that's a blessing. I mean, there's been nothing from the US government for decades. It's been shameful.

I think we need a lot more patient money. We need to train students, taking the biology with the chemistry and the engineering and the physics and building systems and working together in an interdisciplinary fashion—not just to commercialize, but to solve all the problems. To work hard on the problems, put graduate students to work; then those students would come out in the industry. We really need the front end of the discovery and the interdisciplinary research, and that needs to be a little bit more patient than a commercialization thing. It's got to have a 10 to 15 year vision.

CS: That makes sense. It's funny because, as I conduct these interviews, I try to look for common threads—and that is certainly one: let's figure it out before we start. Let's not ready, fire, aim.

GM: Right. If you look at photovoltaics, we've been investing huge amounts in photo science and photovoltaics for decades. If you look at any agricultural thing, there's a huge number of agricultural industries around, universities around, huge industry, huge numbers of jobs already, petroleum refining, you name it, any industry you can think of there is a pretty decent balance between experts in the industry and experts in academia. But if you look at algae, there are no experts in industry because there's not an industry. This will require patience. But the payoff will be enormous.

CS: I have to believe you're right. Thanks so much for your time.

For more information on this contributor, please visit:
http://2greenenergy.com/renewable-energy-facts-fantasies/.

WIND POWER

Clipper Windpower is one of the most visible organizations on Earth in the race to provide solutions that offer utility-scale clean energy. The company strives to advance the technologies and services that make its customers successful in the expansion of wind energy, lessening the impacts of fossil fuel generation. Dr. Amir Mikhail, Clipper's senior vice president of engineering, was good enough to favor me with an interview.

Craig Shields: Thanks so much for your time here this morning. Perhaps we could start with a brief history of wind power.

Amir Mikhail: Well, you know, it's ancient. Hundreds of years ago, the European landscape was dotted with Dutch windmills that were used for pumping water from the land, and making flour from grains—and there're some of them in Holland that are still actually working.

CS: So I understand.

AM: Wind has been utilized across the centuries. For example, to generate power, wind was very prevalent in the Midwest before rural electrification. Before the grid became ubiquitous, farmers in the Midwest would use wind turbines to charge batteries and then use the batteries to run equipment in their workshops. If the wind died down, they would take the batteries to the place downtown to get them charged, but as long as the wind is blowing they would get the charge from the wind turbines.

Then after World War II we had the advent of cheap oil and natural gas, and all that served to put a stop to the use of wind turbines. But in the 1970s we had the oil embargo, and the whole subject of alternative energy came to the fore again.

CS: Well, what happened recently to get you folks off the ground?

AM: After the GE acquisition of Enron's wind business, some people that helped develop the technology of Enron left and came here to Santa Barbara to start Clipper in 2001. I joined them as president of the technology company and then later as senior VP of Clipper.

CS: Let's talk a bit about the technology.

AM: Technologically, it's very simple—simply turning a conductor inside a magnetic field and generating power; it's the inverse principle of a propeller in an aircraft.

CS: All right, but I'm sure there are nuances, right? In other words, what would you say that Clipper does better than anybody else?

AM: We are very well known for innovation because we are a smaller company that deals with big competitors, so we bring innovation. In particular, we started designing this turbine with a lot of new concepts, the most important of which is the aerodynamics and the way you make the system more reliable. It's actually a variable speed system; most earlier machines used to work with fixed speed. In other words, you use an induction generator that pretty much runs at a constant speed with a chip in it to produce 60 Hz power into the grid. So there is no power electronics between the generator and the grid; the generator used to be directly connected to the grid.

CS: What was the downside of that?

AM: The problem with fixed speed is that, first of all, the aerodynamic efficiency is always low; the owner wants to operate at different power levels at different speeds. The other thing is, because they run at fixed speed it has a tendency to have higher loads because, think of something that is like a flywheel. It can accelerate and decelerate which allows it to absorb some of these gusts of wind. So variable speed is a very important concept for modern wind turbines.

The first company to produce a variable speed machine unfortunately went bankrupt in the mid-90s. Now we have a new concept in variable speed technology called "doubly fed induction technology." That technology was instrumental in actually doing variable speed at low cost because at that time power electronics were very costly. So, in order to use these converters, you had to use less power in the converter and that doubly fed concept makes that system a lot easier to utilize because it only handles a third of the total power of the machine. So we developed that at Zond and then it was also developed at the same time in Europe.

CS: Wonderful.

AM: Through all the corporate changes, GE has wound up with a very strong portfolio of patents that they inherited from Zond and others. So when we started Clipper, we realized that we had to do something different. So that's when we came up with a total conversion and permanent magnet generator. We have patents on that, so that was one of the innovations that was brought to the industry.

We are also running a new concept in distributed generation drive-train, which had multiple shaft outputs on the gearbox so you have one input on the low-speed shaft from the rotor to the gearbox and four high-speed outputs on the generator. And when we did this design, it allowed us to make the gearbox more reliable and more compact. We are able to get all the ratios that we need in two stages rather than three stages and made the gearbox more reliable.

CS: That sounds great.

AM: Yes. We have brought in a lot of these technologies. Then in 2003 we received a grant from the Department of Energy to develop a very big machine; at that time the prevalent size of wind machines in the US was 1.5 mW. We actually built a 2.5 mW system—at the time, the largest machine in the US.

CS: What's the diameter of something that's 2.5 megawatts?

AM: It's roughly somewhere between 90 to 100 meters.

CS: Wow.

AM: At different wind speed classes you'd like a different rotor. So we have a machine which is put in very high wind resources that is 89 meters. Then in lower-class winds we have a 93- and a 96-meter product. And we are now developing products at 100, 102, and 104 meters.

CS: Let me ask about where this is going financially. What's the cost per megawatt? In other words, if I say I want a half a gigawatt, what's that going to cost me?

AM: Currently, I think, an installed system costs somewhere around 2 million a megawatt.

CS: That's not bad.

AM: No, it's not bad at all. It's $2000 a kilowatt, but that is system costs, remember; this is not the cost of the turbine. The prevalent price of the turbine is somewhere around $1200-$1300 a kilowatt.

CS: So the rest of it is installation and grid tie and so forth?

AM: Yes, the rest of it is installation, foundation, roads, I mean it's a wind farm, right? So substations, erection, grid tie in, and all that stuff.

CS: I have a friend in New Zealand who has told me about a breakthrough he's made in materials science that he claims enables fabrication of enormous wind turbines. He thinks this is important because the bigger they get, the harder they are to fabricate. Is there any potential truth to this?

AM: I think you're talking about a shrouded turbine.

CS: Yes, that's exactly what I meant.

AM: Yes, a shrouded turbine gives you a higher efficiency because if you put the rotor inside a shroud like a jet engine, that allows more air to come through and allows you to accelerate the wind through the rotor. But the problem has always been the cost of that shroud. And remember, in most horizontal axis machines, you have to yawl that shroud and you have to pitch the blades inside the shroud and so on and so forth. So it's a lot of cost associated with the shroud. So the point is, can you produce a kilowatt-hour cheaper with that shroud than you can with a regular horizontal axis machine? I don't think people have actually been able to prove that yet.

CS: OK. Let me ask you a little bit about the cents per kilowatt-hour.

AM: Wind is somewhere between $.05-$.08 a kilowatt-hour, depending on the wind resources. It's a function of prevalent wind conditions, obviously. If you have 22 mph or something, you get four cents a kilowatt-hour. But if you have something like 13 or 14 mph, you get eight.

CS: OK. Please talk a little bit about the energy storage issue associated with wind, if you would please?

AM: Well, I think the storage issue is, what do you call that, a red herring?

CS: You mean something deliberately to throw people off the track or to confuse people? Yes, that's called a red herring.

AM: That's exactly what I wanted to say. It's a red herring because really wind integration into the grid so far has not been costly. Remember, there are a lot of gas turbines, coal plants, nuclear plants, on the grid. So these plants have to deal with the variability of the load because the load is not constant. When you put the wind resource, you're adding to the variability of the load by adding a variable wind resource. But the grid has been able to handle that so far.

Right now we have 35,000 megawatts of wind power, which is about 1% of the total energy. Storage becomes an issue when you go all the way up to 25, 30, 35%. But below 20- 25% of the total grid capacity, a variable resource can be handled by spinning reserves in order to be able to deal with the variability of the wind resource.

There have been a lot of studies by NREL (National Renewable Energy Laboratory) and by the utilities that show that the additional costs of having to deal with the variability of the wind resource on the grid is less than half a cent a kilowatt-hour. So it's less than 10% of the cost of producing electricity from the wind.

So far it's not an issue. However, in places where wind has proliferated for a long time, in places like Spain, they have 12% or something like that of their electricity coming from wind. Germany, they have 8% of their electricity coming from wind, Denmark has 20% electricity coming from wind. When the load is down and the wind resource is up, in Spain they were producing 53% of the energy for the grid, and it did not cause any problems. We are talking decades away before we really have to deal with storage.

There are certain things that we can do right now which would help the utilization of wind. One of them is expanding and modernizing the grid—you know, bringing in more grid from the Midwest to the east and west coasts. It also needs to be balanced. Right now, every little area within a region is a balancing area; each area has to balance its load and energy source on a very small geographical area. The more you integrate these balancing areas, the more you will have geographical diversity and the more we'll be able to say if the wind was not blowing in point A, maybe it's blowing in B, C, or D.

We also have to make the grid smarter, where the people that are controlling that grid can know at any point in time how much of the power is flowing in and is flowing out; that would give you a lot more control over absorbing a lot more variable resources. We also need to put in smart meters so people know more about their power consumption.

But, ultimately, yes if there is a cheap way of storing power, that would make wind a lot more attractive, but these storage devices are very expensive.

CS: Yes, of course. Let me ask you a little bit about the apparent controversy associated with wind. Some people have been displeased with the wind industry in that it sometimes prefers pristine wilderness, rather than locations where there is already a significant human footprint.

AM: Well, would you rather lop off the top of a mountain to produce coal?

CS: Well that's exactly how I responded when I first heard this complaint. It seems like a completely ridiculous thing to say. We'd rather have gunk in the atmosphere or another Chernobyl?

AM: Right. You're probably following the example of the mountain in West Virginia, where people in this area chose to actually destroy that mountain to produce coal, rather than put a wind turbine on it that they can move off later. I don't understand that. You have to choose. What is the alternative? Put a nuclear power plant up, or put in a coal plant or a gas plant?

The point is, there are choices. Each one of these choices has a certain kind of damage associated with them. Nuclear, nobody has the space for a place to store all the spent fuel rods. All these methods that allow you to minimize the storage volume allow you to produce nuclear bombs. Right? It gets you a fissionable material.

It's true that the industry has made some mistakes earlier on, like where they put wind turbines in an area where there's a lot of raptors; that was not good. Now what's happening is that technology is improving and turbines are getting much larger, so some of these birds now are able to navigate those more. There's also some research being done on bats and wind turbines. And apparently bats are being affected by wind turbines because their little lungs cannot handle the pressure differential that is caused by the spinning rotor in the air. So, there is a lot of research needed. We're spending a lot of money on bat research trying to make sure that we do everything possible to mitigate damage. Or we shut off the turbines at a time when the bats are foraging, which is normally a very low wind area because bats normally don't fly when it's very windy.

CS: In closing, I wonder if you could comment on competitive technologies like solar thermal and molten salt if you don't mind.

AM: It's a fantastic resource. I mean it's cheaper than photovoltaics actually. But PV is very sexy and people love to have something that's not moving on their rooftops that produces power. But in terms of mass production of energy, solar thermal with its reflecting mirrors, parabolic dishes, parabolic troughs, all these are proven to actually produce energy. But you have to have solar resource. Those are very limited to the southwest because there is a lot of land and a lot of desert and you have to deal with abrasion from sandstorms and all that stuff. So that adds into the cost of the long-term. So the cheapest renewable resource is wind—especially if you are located in windy areas. You can have electricity at four or five cents a kilowatt-hour all day.

CS: Great. Well, Dr. Mikail, I can't thank you enough for this.
AM: Absolutely.

For more information on this contributor, please visit:
http://2greenenergy.com/renewable-energy-facts-fantasies/.

SOLAR ENERGY—
PHOTOVOLTAICS

I'm proud to call Bruce Allen a friend of mine. In addition to his abilities as a tennis player and great chef, he functions as 2GreenEnergy's senior consultant in solar energy, specializing in large-scale solar electricity projects. He has written numerous articles on solar technologies, solar energy systems, and national energy policy.

His recent book Reaching the Solar Tipping Point describes the key technologies and applications that are enabling solar energy to become a primary cost-effective energy source. He has designed solar concentrator systems sold worldwide and worked at the Jet Propulsion Laboratory, under contract to NASA, DOD and the US Missile Defense Agency.

Craig Shields: Perhaps we can start with a few of the basics regarding photovoltaics.

Bruce Allen: It's really simple. Originally, Albert Einstein discovered the photoelectric effect. That's what he won the Nobel Prize for.

CS: Really? I thought that was Max Planck.

BA: No, actually the photoelectric effect was discovered by Albert Einstein, oddly enough. I believe it was in 1904, and then in 1920 he actually got the Nobel Prize for it. Light hits a semiconductor, kicks an electron off, creating an electric field sweeping the electrons away, which creates a hole that is swept in the opposite direction, which creates a circuit. The challenge is to make these devices as efficient and as cheap as possible. Conventional photovoltaics are based on silicon; you dope silicon.

CS: When you say "doping," you mean you put boron on one side and phosphorous on the other, creating a force that causes relatively positive and relatively negative things to move.

BA: That's right. That's about 85% of the market right now, maybe 90% of the market. There is conventional bulk silicon, which means the silicon layers are relatively thick. The thin film silicon is developing pretty rapidly, led by Applied Materials, the big player in making the machinery that efficiently makes large panels of thin film silicon.

CS: OK. Not to dwell too much on the physics, but what is the difference between thick and thin film from a standpoint of what's actually happening?

BA: The main thing is the thickness of the actual active layer. Traditional photovoltaics were made by growing bulk silicon crystals in ingots. These ingots can weigh many kilograms, then they're sliced into wafers. They're getting down now to a hundred and sixty microns, more or less—and then those wafers are polished, doped, and so anyway these wafers are relatively thick.

Thin film is vapor deposition; you're getting layers very, very thin, using compounds like cadmium telluride. The largest maker of thin film silicon equipment is Applied Materials; they're the world's biggest semiconductor equipment maker. The other major player is First Solar—together they probably have 95% of the market of cadmium telluride.

There are other semiconductors, for instance CIGS or cadmium indium gallium di-selenide, which is a somewhat more complex and harder-to-make thin film technology, but is actually now starting to be commercially produced. A company called Solyndra has gotten about $800 million from the government to develop this further. That's a lot of money to ramp it up. Now it's still a question of how successful they'll be, but clearly people believe that they'll be successful, and that's another thin-film technology.

NanoSolar is doing the same thing with the CIGS. Their approach, though, is essentially nanotechnology applied as low temperature paints. At least that's the theory. They don't talk much about their technology but that's what people believe they're doing. And so it can be done at low temperature almost like sprayed on paints. The traditional way for thin film is very high vapor deposition; you're talking about hundreds of degrees.

The main difference between the technologies is efficiency. Bulk silicon theoretically gets up to around 32-33%. Sun Power, I believe, has achieved commercial quantities of 23% efficient wafers in bulk silicon. With their panels they've just now achieved 20% panel efficiency, which is pretty good. Most of the rest of the bulk silicon industry is in the high teens. With the thin film silicon you get up to 8-10% now. With cadmium telluride, I believe they're shooting for 11% next year. Ultimately they think they can get to 16% or so. CIGS, theoretically, could get up to maybe 20%. In the lab I think they've done 19%; I believe the best commercial stuff is about 10 or 11% also.

CS: And solar-thermal is in the same range, right? 17-20% of total incident sunlight.

BA: Yes, but with one exception. That's the Stirling Engine, which is about 32-34%.

CS: Would you digress for a second and explain that?

BA: It's basically heating the gasses and running a piston. There are two main reasons the efficiency is so much higher for them. First of all it's a circular parabolic focus, so you're getting high concentration. You remember from thermodynamics that the higher the energy difference, the higher the efficiency. So energy efficiency is driven by the temperature you can achieve with the focusing of the light from the solar-thermal. So the Stirling system has very high focus concentration.

Coupled with this is the fact that the Stirling Engine is at the focus, so there's virtually no heat loss in terms of sending the power from the tubes to the power plant.

CS: So it becomes kinetic energy—and then electrical energy—right there at the focal point?

BA: That's correct. These Stirling systems right at the focus of the reflectors that are twelve to sixteen feet in diameter—each one has its own engine at the focus, and they're running the Stirling Engine and it's generating the electricity right there. Those are in the mid-30s for efficiency and there are some very large contracts for them to be putting them in the desert. I think it's very questionable long-term how well they'll do commercially, because I haven't seen long-term data on how well they've been doing in large deployments, whether there are mechanical issues, and so forth. The solar thermal traditional trough and towers have been studied and run for decades. So they have good data on those, where the Stirling Engines have been studied for much less time.

CS: Well, let me go back to PV for a moment. The fundamental problem with the efficiency is rooted in the fact that in the incident sunlight comes with stuff that you can't use as part of the electromagnetic spectrum. You get more energy in a photon than is required to release the electron, or you don't get enough, or whatever.

BA: That's right.

CS: So you get these efficiencies in the high teens so far. Can you tell me about quantum wells or quantum dots? Are these attempts to deal with this?

BA: Yes, again it's like going from conventional Newtonian mechanics to relativity. The world changes. That's how quantum effects creep into it. In bulk silicon, you get this energy band gap. If you're above the band gap you kick off an electron, or can, and then any extra energy is wasted; if it's below it, you don't kick off anything and it basically is wasted as heat if the photon is absorbed. With quantum dots, what can happen is the photon is actually absorbed by a quantum crystalline structure, which essentially starts resonating and rather than only one electron being kicked off, it's capable of kicking off more than one electron. So immediately you see the advantages of that. You can now capture a higher percentage of the photon's energy because if you can kick off multiple electrons at different energy levels, you can essentially capture more of that photon's energy.

It's very preliminary in the labs. I mean there's very, very basic research going on with it. There's reason to believe you can get efficiencies between forty and sixty percent. And the number of materials is huge that you could try to do this from and there are all kinds of complications in doing it efficiently, and cost effectively. But in the long-term, that's probably where the whole photovoltaics industry ultimately will go is these quantum materials.

CS: OK. Let's talk about PV at the homeowner level, and at the utility level.

BA: Well there is utility-scale PV already deployed. California has utility-scale PV, but we have to make some distinctions. Germany has been the largest developer of utility-scale PV, because of the heavy subsidies and feed-in tariffs that they've done—even though the climate is not at all ideal for that. But Germany has what I consider utility-scale, multi-megawatt projects that feed the grid for the sole purpose of providing the grid electricity. So those would be defined as utility-scale.

Germany guarantees 65 cents a kilowatt–hour and there's a sliding scale over time and the number of providers and the amount of power, it's eventually, a feed-in tariff starts very high and then gets smaller.

CS: 65 cents? Isn't that huge? Our national average is like 10 cents at the retail level.

BA: Right, but in Europe electricity prices are much higher anyway. Say if you did your project in 2004, you might have been guaranteed 65 cents for X number of years. But as the industry scales up and you start getting many megawatts and ultimately gigawatts supplied into the German grid, 2009 comes along and maybe now the government says

"Ok, going forward we're only guaranteeing 40 cents per kilowatt-hour," and in 2012 it may be less. And so it goes down over time, but generally the early providers get guaranteed rates for enough time to recoup their investment. That's the goal—to stimulate the investment so that investors can have confidence that they'll make money on their project.

Right now, you know you also have utility-scale projects in California. Sunpower is doing them, so you have utility-scale projects both in photovoltaics and solar-thermal. Photovoltaics is clearly far more advanced, but the solar-thermal projects tend to be very large scale, or will be. It's very early in the life cycle of solar-thermal in terms of large-scale commercial production. Spain is probably a leader of that as well. In southern Spain they have both solar-thermal farms and photovoltaic farms, and they're growing rapidly. I know Korea has some large projects that have been done and are ongoing. China now is looking at some very huge utility scale photovoltaic projects.

CS: Let me ask you about energy storage. It seems to me that all technologies have some issue here. Obviously if you have a coal-fired power plant that's cranking at exactly the same power output 24 hours a day, you have a problem because you've got extra off-peak power. With solar, you've got an issue because the sun isn't shining 24 hours a day.

BA: Right. Coal power plants tend to like to run without interruption. You can, to some extent, throttle them up or down within certain practical limits, but they do tend to want to be run continuously. That's why electricity at wholesale rates is cheaper at night—because the vast majority of our electricity comes from these continuous sources. That's why California's electricity at wholesale level can be fifty percent or less the cost at night as opposed to peak. In peak summertime, rates can be four or five times more. Running coal power plants and just conventional fossil fuel power plants and hydro and all that sort of stuff, you don't need to store it obviously; it's continuously available. Energy storage becomes a big deal to renewable energy, specifically solar and wind, because although in the desert solar is pretty reliable, you know, 90% of the time or more—but the sun doesn't shine at night. So there you've got an issue of primarily storage at night, and a few days a year that you've got clouds.

Wind is much less predictable and if you want to store electricity from wind energy, you might have to do that any time of the year for significant periods of time because it's much less reliable. So, you have the basic technologies...I mean there are a number of ones that they're investigating, but fundamentally it's done through batteries, or would be. There have also been some studies done on storing electricity energy as compressed air in caverns and running turbines from the compressed air later on. There was a paper in Scientific America a couple years ago that talked about that approach with photovoltaics. In theory you could do that with any electrical source. You know, taking electricity and converting it to compressed air in caverns and then running it out through the turbines again, it doesn't really matter where the electricity came from. There's a lot of work going on in battery storage, but it's nowhere near commercially practical right now.

CS: I would think not. If a lithium-ion battery pack in a single electric vehicle is a $15,000 item, I would think that it would be a seriously poor decision to do this at a utility-scale.

BA: Right, and even if you chop the cost down by a factor of 10, you're still not even close to being really practical, because it's just enormous amounts of power that need to be stored. So what the utilities are focusing on is power regulation and interruption. There you can have megawatt class batteries which, for a few minutes or an hour or so, are very practical to use because you want to regulate—or level—the grid. You know, you want to smooth out supply disruptions—short-term storage for load leveling. In those cases, and that's what the main focus has been recently, in terms of the batteries, is to do that. Utilities are giving contracts to some of these battery providers for relatively small, modest, megawatt-class batteries to do those kind of grid applications. The government has grants and incentives to help them out with all that kind of stuff. But in terms of large scale storage, it's probably going to be a very long time before you're looking at practical storage technologies for truly, regional level electric storage.

CS: In our past conversations we've talked about molten salt. If you're implementing solar thermal, you start with heat energy; I would think that would be a huge advantage.

BA: Right, solar-thermal has a much different and much more efficient and cost-effective way, potentially, of storing heat energy that's generated. So before you generate the electricity, you can essentially store it into large molten salt tanks. I've talked about this in my book (Reaching the Solar Tipping Point), and that's exactly how it's done. And there's a facility operating now in Spain that's demonstrating this on a commercial scale.

CS: I guess when you look at the curve of peoples' energy use, ebbing at three or four o'clock in the morning, accelerating at the breakfast hour, and then continuing, peaking maybe at six o'clock at night?

BA: Late afternoon, depending upon where you are in the time of year and all that sort of stuff. When you're talking late afternoon you tend to get peak, or early evening, again depending upon the season and where you are.

Renewables will grow in percentage of total use in the electric grid. And over time it'll move from purely as available either in the daytime for solar or whenever the wind's blowing for wind. But as these percentages keep rising, ultimately at some point, if the percentage continues to grow, you would reach a point where you start reaching the storage issues. But that won't happen for a very long time because renewables supply such a small percentage, at least in the US. In certain European countries, renewables supply the majority of electricity, but those tend to be hydroelectric and that's very reliable.

CS: Bruce, this has been fantastic. Thanks so much.

SOLAR THERMAL— CONCENTRATING SOLAR POWER

My main hope about publishing this interview is that I've somehow concealed my belief that CSP - concentrating solar power—or "solar thermal" more broadly—will soon dominate the world stage as the renewable energy technology of choice. I was thrilled when Dr. Mills accepted my invitation, as he has been known worldwide for pioneering Compact Linear Fresnel Reflector (CLFR) technology and for his work over the past 30 years in non-imaging optics, solar thermal energy, and PV systems.

Craig Shields: I have to say that I go into this interview favoring solar thermal, based on a lot of material I've read. I've been looking forward to speaking with you for some time.

David Mills: Oh, that's very nice to hear. It's an interesting field.

CS: Let's start with the notion of scale. I think we use 5.4 trillion Watt-hours of electrical energy in the United States every year. What's the trajectory of getting solar thermal to a point where it can make a significant contribution?

DM: Oh, we can certainly do that. It's a little bit difficult to tell what the ultimate distribution of different technologies will be, but I think it's probably going to be a little bit different from what most people think it is. However if you have technologies with a big resource like wind or solar, for example, they can all scale to very large. Each of them has the capability to take on the entire electricity load. But the question is how much does that cost and do they do it in a way where we have reliable energy?

CS: Yes. Of course there are numerous sub-topics on that in terms of base load versus peak, the fact that the sun shines during the day, molten salt energy storage and so forth. But here's a question: do we really need to develop several different technologies at the same time?

DM: I think you'll find that these technologies work together better than they do apart. Fortunately, though, there are some very strong correlation between what solar does and what humanity does. That goes way back to the fact that we are animals that have evolved to come out in the daytime when the sun rises and go to sleep when the sun goes to bed as well, and therefore our activity is correlated to some extent to the sun's activities. Not completely, of course; we live in an industrial society and there are things that run all night, but there is a strong correlation there. And there are other correlations as well.

The second is that we do find, though this is very early in a research sense, there is a kind of anti-correlation between the presence of wind and sun which is very beneficial.

Finally we have the possibility of storage, which I think is going to be a key technology that allows the blending of those to give you that completely reliable system. We are developing this storage system but we have been delayed by a number of things. I think all of these solar companies have had a difficult time in the past year, which has cut back on the research we've been able to do.

In addition to the availability of funding, the game keeps changing. You can try to develop, say, a technology for a certain temperature range, and then other technology advances so quickly that we're not working in the right temperature range anymore. Things are moving so spectacularly quickly that it's hard to get your feet. I see things moving now quite quickly toward higher temperature in centralized plants—over 500°C for major technologies. And that gives you efficiencies and cost that are much closer to those of conventional generation.

CS: That's good to hear. A friend of mine said recently, "Look, there's plenty of clean energy out there. The only real question is what are we willing to pay for it."

DM: He's right.

CS: Well that's why I asked you the question of scale. One would assume that since solar thermal invokes extremely common and inexpensive materials, one would think this would scale better than almost anything that you could name.

DM: I think so. Even with batteries, for example, the amount of lithium in the world is a question, or if you're looking for solar with cadmium telluride, while both cadmium and tellurium are fairly rare materials. You would very easily run out under the current production methods if you're doing a massive expansion. What we do is glass and steel basically, and some concrete for footings and things, but mostly very common and widely used materials. Probably the thing that will expand the most is the actual glass we use; you would use more glass than in a typical construction site, so it's good for the glass industry.

CS: Did I read somewhere that you've changed your business model?

DM: Yes. Originally we were going to build large projects ourselves, due to legislation that happened to be in place—a 30% income tax credit that was a big deal in terms of competitiveness; it was not allowed to be used by conventional utilities. In other words you could use it if you were an IPP, an independent power producer, but not a utility. And secondly we had to prove the technology, so it was best to build our own projects.

But getting money on the scale for projects we would have to pay interest at an exorbitant rate. What happened last October a year ago was that the Congress put through the ITC provisions, extending them,

I think, for seven years. In the fine print it said that utilities were able to utilize the ITC too. So we felt that we didn't have to do the projects, we only have to supply equipment to them.

CS: I see, and are you happy with that?

DM: We're very happy. We were the first to announce this change in model, and everybody said oh, well, they're collapsing, they won't be able to handle big projects anymore, so the move was misconstrued. We are going to build very big projects but were going to build the solar field. That's physically the much bigger part of the project. In terms of money it'll be around half of a typical project.

But we wouldn't be building the power block where the turbines are— all those other things which are actually very traditional technology that other people can do quite well already, and that's not our expertise. This way we get income right away we don't have to raise huge amounts of capital; the project developer or utility raises that.

CS: All right. Well I hope that works out for you.

Are solar plants terrorist threats? Someone wrote that in response to one of my blog posts. I would think they could be distributed in much the same way anything can.

DM: Yes, that's right. I suppose this is the usual range of stuff that comes out of the nether regions of the existing industry. In terms of security, I believe there is a strong argument that this is much more secure. In the first place, you're producing all of your energy for the country inside the country. All of the electricity in the United States can be produced in the United States, indefinitely. There's no shortage of fuel.

Now you could say that could be done with coal, and that's true, but we can actually go further than that. If we electrified the economy and say that we have to have clean energy sources in order to make those comply with the climate restrictions, we are saving a lot of oil imports as well.

We will have generation as we do now, but we've got to look to expanding the electricity component in our society, not only into stationary applications as we have now, but into areas that are using heat, for example into the home. For example if you have a home now that is heated by oil, you might heat that using a reverse cycle air conditioner in the future.

CS: And you have the transportation sector, of course.

DM: Yes, electrifying the vehicles. I've got a plug-in hybrid Prius. I commute every day, mostly on electricity. I know it's possible because I'm doing it. Those things make huge differences, because once you connect those big loads to the grid, you now have access to all of the renewable energy in electrical from all around the country. It's a very important step to take.

CS: I agree. You probably saw that $3.5 billion in stimulus money for Smart Grid that was just awarded this week. And that's obviously good for the idea of high voltage DC that means, ultimately, that the physical location the power was generated becomes less and less relevant.

DM: That's exactly correct. I try to make the analogy that after the war they developed the national highways program, which is set up as a federal system—the backbone of the road system. Then from that backbone you have feeder roads which are owned by the states. It seems to me that you can do an exactly analogous thing here, the federal government can take control of the backbone, which facilitates their carbon targets and the states retain control of the local sectors as they do now. And I would say that the whole thing would pay for itself, because you would start charging for the electricity going coast-to-coast. You know, if a kilowatt-hour goes from A to B, it would attract a transit payment. So as long as the government puts out the initial money, we can pay it off in 15 to 20 years or whatever; I don't think it would be a big impasse. We have to do it anyway, because the grid's a mess anyhow and people have been talking about replacing the grid just because it's old, so let's build it right.

CS: It sounds like it's not too bad being David Mills. It sounds like you got the right thing at the right time. Yet there must be one or more significant challenges that you face. Can you talk to that please?

DM: Sure. We are a start-up and all start-ups face difficulties. We're a mouse among the dinosaurs, trying to move in. We have to live with that, and that's probably, on the day-to-day, the hardest thing. It's nothing to do with the design of the technology. If you have a new technology, it's risky and therefore people charge you more for the interest that's associated with a big project. The interest rate can drive you to the point where you're not economic with your competitors. So even though you might be 20% cheaper, you might be charged 20% more for your money. So you end up with a technology that costs the same that is new and risky.

CS: Right, I'm with you. Everybody thinks when they see the Kleiner Perkins name that money's no object. But that's not funding these gigawatt efforts. That's administrative; that's getting the corporation kind of on its feet, right?

DM: Yes, that's right, it's the initial funding you have. You start up with the initial investors before Kleiner, and then you come into Kleiner and Khosla or other interests that we have. We have five big investors, actually. We were moving out of that stage and on into the next stage facing all solar companies, and that's finding a big partner basically to go with. The reason for that is that you have to have the resources to undertake a lot of engineering, a lot of big plans, but also you have to guarantee the work that you supply. People will not buy from you if you can't guarantee your work.

It's pretty hard to bootstrap yourself in that position where you can satisfy the warranties from zero money.

In the end it comes down to this. You have to find welcoming traditional industries. It turns out that there are those industries out there. They're not the utilities. The utilities are all right, by the way, I'm not damning the utilities at all; they just turn heat into electricity and this is another way of doing that, so it's not beyond their understanding at all. It's not a big jump for them to take. It's just the boiler is different.

In fact, we characterize ourselves as a solar boiler company. It sounds a little bit mundane, but when you look at what you're doing it's actually very similar to Foster Wheeler or any of the big boiler companies. We're trying to create the initial energy for the turbines. The rest of it is all the same, really.

CS: What type of partner do you think would be best for you?

DM: A strategic partner—a partner that offers more than just money for investment. It could be a project developer, it could be a manufacturer of turbines, it could be someone that does the EPC (engineering, procurement, and construction), which means that they actually manage the construction on the project, making sure it's all finished off and done in the right way. There are some companies that do all those things. We are in discussions with companies of that nature.

Solar thermal is much closer to the things in their portfolio, where PV is quite different. The other projects they produce may use turbines, and PV doesn't. There's a level of learning to PV, and PV companies can actually work their way up organically much more because you can start small and you can make the next array a little bit bigger and grow them fairly fast and organically, especially since there have been good subsidies.

Our technology has the great advantage of being steadier in output, but has the disadvantage of being very large. That means it's really hard to grow organically. You have to go to a very big project right from the beginning, otherwise you're very low in efficiency or the project isn't worth the engineering effort that goes on.

So we're playing a different game, but the other difficulty is that we're fighting on several fronts. We're obviously the target of conventional energy companies who put out misinformation about our technology, that sort of thing. A common one is that solar can't run a base load.

CS: Yes, exactly. That's the thing that really irritates me.

DM: Yes, it irritates me too, because, since, I think it's1993 we've been talking about this problem. Base load is the thing that nuclear power plants and coal plants produce, but it isn't what we want. You can tell that it's not what we want because we have off-peak rates to drive usage into times of day that wouldn't normally use much power. As I've shown in a paper or two, solar produces a much closer correlation to human activity, and that means that if we had a solar economy, per se, we probably wouldn't have off-peak rates. Because you could store a little power overnight, it would correspond very closely with what we need.

So that base load is not something we want to produce. We can produce base load—and there is a plant that does pretty much that going up in Spain right now; it's a tower plant with 15 hours of storage. That plant has a projected capacity factor of 75% and it's going to be finished in 2011.

That will show that it's possible, but I actually think it's the wrong path. We need storage but we don't need as much storage. I'm actually doing some work which will come out later this year with some others, partly at a University, and I think it shows that if you have a fair amount of wind and solar in the system, the storage you need in the solar component is relatively small. That's going to be a groundbreaking outcome. I can assure you that this is not something that you can read about right now. Nobody's ever done this type of analysis. We analyze the output of the United States on an hour by hour basis over a whole year.

CS: Wow, that's interesting. When I met Ray Lane at a conference in Detroit last fall, he really took the industry to task; he was not too happy with these people. He said words to the effect that rhetorical platitudes do not get the job done. He noted that we dominate in IT, but five of the six great renewables companies are outside the United States. We cannot afford to not make these investments and to not win this game.

DM: Well, that's right. The United States was leading in practically all of that in the 70s. A huge amount of work was done in the 70s. It was all shut down quite deliberately by Reagan. The first thing he did was take down the PV array on the roof of the White House—and the second thing he did was shut down half the Solar Energy Research Institute. He did that in very short order—within a day or two.

I think that decision cost the United States a huge amount because the lead drifted off to Europe, I think in large part, but even Australia was doing more in many areas than the United States. The budget was so suppressed that it was just very difficult to get anything up here. The only thing that did proceed was wind. Wind had a lot of research money pouring in—especially in Europe where it had a market which still had some subsidies in place.

In the mid-1990s it was very interesting watching the takeoff of the technologies; in 1995 we got wind shooting ahead, and then you had PV, which since about 2005 has taken off. I think in 2006 or 2007 it was growing in 50% per year, which is colossal. Now it's flattened out but it'll probably go up again as soon as the economy recovers.

We think solar thermal is not far away with the inflection point. It's very close because right now around the world there are 22,000 megawatts of projects being planned. We think that there was kind of a little informal think tank at a recent conference and they determined that about 14,000 of that is real—14 gigawatts that is actually ready to pull in. That's not nearly as big as wind, but it's clearly not small beans anymore.

Once you get things like that happening, the industry starts to self-invest; it doesn't depend on VCs (venture capitalists) or government funding so much anymore. The wind industry does most of its investments internally now.

CS: What do you see as the gating factors?

DM: The technology is still a little bit young in the sense that it is going through a period of rapid development, but that's happening very quickly and the underlying collector technology can be implemented right away.

CS: Well, do you mind talking about that a little bit from a technological standpoint? What needs to happen before we can have gigawatts of this stuff?

DM: We have to bring the cost down—and there are a number of ways of doing this. We have right now two basic groupings of systems and then two sub-groupings, if you will, of each. There are the high concentration systems, one is a dish type and one is the tower type, and the tower type is really what we call a Fresnel analog of a dish collector— so many smaller mirrors acting together like a dish. In the linear systems we have the parabolic trough and the Fresnel analog, which is our linear Fresnel system. So there are these four stars in the constellation. The linear systems have lower temperatures but are cheaper to put down in the field for a couple reasons. We think the Fresnel ones in each case are the cheaper of the two, cheaper than moving round large mirrors on huge structures. You'll see a lot of parabolic trough plants going in because that's the technology that was proven over the last 20 years and therefore it's the most developed and understood, and its development is almost stopping because it's almost optimized.

CS: I see. Is it useful to characterize these in terms of a percent efficiency? Obviously there are other losses associated with solar thermal to bring it to what it is ultimately—is that 18 to 20%? But I wouldn't think a lot of them are in the reflective nature of the parabolic trough itself.

DM: No, all of these collector types use fairly similar quality mirror. And that's been gradually improving over the last two or three years. We are getting from about 93% to very close to 95% in the most recent ones.

CS: OK, so where does most of the loss occur? In the creation of the steam in the heat exchanger? The conversion of heat to mechanical to electrical energy?

DM: The single major loss, if you will, is the turning of heat into electricity through a normal heat engine, which will be a steam engine and steam turbine in most cases, but also could be a combined cycle plant and high temperature units.

But there's a secondary major type of field losses. When you define your efficiency, you have to be very careful what you are defining, and in fact it proves very difficult to find a definition of efficiency that fits all of those four possibilities. For example, the parabolic trough defines its efficiency in two ways. One is peak efficiency when the sun is overhead there. Then there's an annual efficiency which takes account of the movement of the sun and what we call the cosine losses. It might be coming in at an angle therefore your temperature is decreased and you have less output for that reason. That is a substantially lower figure, than if the sun were facing the collector at all times. You say, "Well of course the sun's never going to be facing the collector at all times," but in some cases it is. With parabolic dishes, for example, that faces the sun most of the time. And therefore it is the least of the cosine loss. However it's the most expensive to build.

CS: I see. Yikes, this isn't a straightforward number then, is it?

DM: No. You have people developing different approaches—some vastly more favored than others. For instance, very few people are developing dish type concentrators for solar thermal, since there are many practical mechanical issues with them. There's a group in Australia that's been doing that for some years and is trying to develop that. But, you know, in general, my opinion is that it's very hard to get a large-scale manufacturing enterprise that makes sense in terms of the installation and the maintenance of such things.

Also they are so large that you can't get people in there to clean the mirrors. How are you going to clean if people can't reach it? Certain collectors need cranes to clean them - very expensive. There are thousands of these dishes in a proper field, because even though they're big objects individually, to get a very large area you have to have many of them.

So what you have to look at is something that collects energy cheaply. Now, you might say that "okay the linear system should be defined in the same way, they both lie on the ground, they're both flat..." but actually they're different too. The parabolic trough maintains a geometrical shape at all times. Its absorber is in the same position relative to its mirror. The whole thing turns around in one axis and faces the sun.

In the linear Fresnel that's not true. The mirrors move in individual rows, and the collector has a different shape at different times of day—you have to look at the collector as a whole. The absorber doesn't move, the pipe is fixed. So, when you try to define the collection aperture of that it's actually quite difficult because it actually changes during the day. And so trying to define efficiency of that is a difficult thing.

It's been a subject for years at the conferences and I believe it should be done on the basis of mirror area—how much mirror area you use. Mostly, people are increasingly just going by the internal rate of return on the whole thing—some sort of economic cost per kilowatt-hour.

CS: Well, I can understand that—given the complexities you just mentioned. In fact, that was my next question. Where are we as an industry with respect to dollars per kilowatt-hour?

DM: Well, everyone wants to know the answer to this, but it's one of the hardest things to say, because the solar thermal industry is very sensitive to fluctuations in the prices of materials. In particular, it uses steel. It's not very sensitive to the glass price, surprisingly, because the glass now is cheap enough that it's a very minor part of the cost of the system. It's actually the rest of the system that supports the mirror in a very accurate way, and this structure has to have a very good shape to force the mirrors into a very good shape so they will perform well. So it's really the cost of that support.

In the case of big LFRs (linear Fresnel reflectors), we believe there's a lot yet to do in terms of bringing down costs. We have an idea of where that will go—let's just say that the LFR is going to be quite a different animal than it has been in the past. I guess if I had to pick out a general trend, it's that both the linear systems, the parabolic trough and the linear Fresnel system are now aiming at higher temperatures than before. The rationale behind that is just to run a turbine at higher efficiency.

CS: Yes, that's what I understand, because of the laws of thermodynamics.

DM: That's right. The fixed plumbing becomes quite important at high temperatures. You have high temperatures and pressures of steam moving about your system, or you have molten salt moving around your system. It's better not to have moving pipes in such a system. It's just very hard to try to get a high pressure above, say, about 80 atmospheres in a system with moving joints. Yet really high temperature systems will operate anywhere up to 200 atmospheres.

CS: OK, interesting. Well before we leave this subject of dollars per megawatt, I wonder if we can't say something about it insofar as if you talk to the people in the wind or in geothermal, I'm sure they have similar problems associated with quantifying this thing down to the penny.

DM: Wind is probably a decade ahead of solar thermal on the development curve; everyone pretty much understands the design of a wind generator. And with great confidence they give a cost per picowatt according to the current NREL thinking—a long-term price of land-based wind—probably be about 6 1/2 cents a kilowatt hour or something of that order. I think most people would agree with that.

There are positives and negatives in wind. You get more and more efficient as you make little tweaks to the system over time, but then again your good sites get used up. So you start after going to lower and lower wind sites and of these two effects work in opposite ways in trying to predict the prices; it becomes a bit uncertain, but the US government and NREL have done a very large study and I think about 6 1/2 cents a kilowatt hour is about right.

On top of that has to be the cost of dispatchability—the cost of making it a reliable source.

CS: Yes, indeed. Not to beat a dead horse, but this solar thermal industry—including your competitors, Acciona, Abengoa and so forth—they don't ascribe a cost per kilowatt hour?

DM: Not usually, except in general terms. They might have a general study of the future and say "Oh, the cost is going to go down to X," but that's where we run into difficulty again. If we have a fluctuating steel price—and we've see fluctuations on a factor of two in recent years—

and that's the major structural component in your system, it has a huge effect on your cost. The other thing is the interest rate. Wind is an older technology, which means that you can borrow money for your project at maybe 6% to 7%—8%, at the most eight, for your project. If you have a new technology of any sort, it's up around 10% to 12%. That makes a huge difference in the cost.

CS: Oh, I would imagine. I mean, I have a lot of respect for Ray Lane, but I would think that the presence of Kleiner Perkins—I mean they want to make money on this. They're not doing this as a hobby.

DM: Oh, absolutely not. They will do well out of our company and they will make a profit as they intended to do, but the development of the technology requires more than VCs. You can actually contemplate a software company or to even some extent a PV company growing organically from a start and then going to IPO and going through and being its own master. But in this business that we are in that is virtually impossible because the projects are so large. It looks like the average size of the projects will be, I would say, between 200 and 500 megawatts normally. Then often a project is several units like that. So you're talking about billions of dollars per project, and for small companies to try to raise that kind of capital for these projects is almost impossible. So we're all moving toward strategic partners—large companies which can actually provide the backstop of a low interest rate. Also they provide warranty protection for those who use the products, which a small company can't provide.

And so, the industry is doing that right now. I mean, in about a year you'll probably find few or zero companies available for sale with new technology. This happened in wind a few years ago—you saw GE and other big companies all buy into wind, and you'll find that they're the companies now running most of it. There is an organically grown company that supposed to be independent, but mostly you're getting now big utility oriented companies now producing these things.

So, I think that if you look at the field, the field is going that way. I think PV companies, for a long time, have been installing their own stuff, but those are much smaller scale projects.

CS: We haven't talked too much about the public sector in all this. Is there any involvement of, for instance, the DOE and NREL and the stimulus package and so forth and so on?

DM: Yes there is. The DOE has put out a number of packages, for example, just looking for interesting projects. There was one where they were soliciting new storage ideas, and then ARPA-E on innovative energy technologies.

This could change everything. We actually got to the second round in that, but didn't make it through the second round. It was very difficult because the first round cut off was over 90%. You're competing with all of the energy technologies, not just the solar technologies.

I think that the government itself was very surprised at that response. They had thousands of applications. I would be surprised if they didn't think it over and start to try again on a broader scale—with more money than they put in before.

CS: Interesting. I also wonder how fair this is. It seems to me when I look at ARPA-E as an example, I'm kind of disappointed in the results that I see coming out. I can't look into the black box and see the decision-making process, all I can see is the output of it, but I'm kind of disappointed with the companies they select. It seems to me that a lot of these things are bureaucratic, they tend to be slow moving, they tend to be not at all innovative.

Let me ask you about regulation. A number of the people I've interviewed suggest that we have over-regulated in the US—and especially in California. And I read that the oil companies hire stooges to show up at meetings to complain that you're using too much water and they can't grow crops?

DM: It certainly used to go on like that in the 80s. There was always the fossil fuel person on the public committee to slow things down; that is common knowledge. Recently, we also had some problems with the unions, because we find that there is an opportunity in the permissions process for public consultation and this can be abused. And it turns out that companies on a particular project that were using labor had no such problem, but companies that were not using union labor were having a terrible problem with permissions, because they're getting a lot of letters and requests for information.

CS: Well let me ask you this: How can I help?

DM: Carrying the message forward as you are doing now is a big thing. Trying to break down the big propaganda regarding base load is very important. If your blog is widely read, and I think it is, then that's one of those things that really helps a lot. The public recognition factor is a big thing. Decision-makers in the country don't understand the great superhighway concept very well yet. They don't understand that solar can be an array about 150 miles x 150 miles in cumulative size and powering the vehicle sector and the electrical sector. They don't understand that wind and solar together look like they can just about do the whole job.

CS: So we do have a job to do here.

DM: It is a big job—and I think it just takes years and years and years to drum it in. Of course we have the anti-propaganda the whole time and green wash coming from the oil companies and the coal companies. And it's hard to match those budgets. You've got, what, seven lobbyists for every member of Congress or something like that?

CS: Yes. It's awful.

DM: My industry can't afford that kind of stuff. So anything the blogs can do for free is really good for us. So we don't mind keeping in contact and trying to do that.

CS: Well, I'll do what I can; I can promise you that. It's been my pleasure speaking with you.

DM: Glad to do it.

For more information on this contributor, please visit:
http://2greenenergy.com/renewable-energy-facts-fantasies/.

GEOTHERMAL

For some reason, geothermal energy is often overlooked in our discussions of renewable energy—and I've been guilty of this omission myself. In the panel discussion I moderated at last falls Alt Car Expo on charging infrastructure, the very first guy to ask a question of my panelists and me lighted into us for never mentioning the subject of geothermal once in our hour-long conversation. It's a mistake I'll not make again.

Here is a discussion with Paul Thomsen, spokesperson for Ormat Technologies—whose geothermal power plants are a field-proven, mature commercial product operating worldwide.

Paul Thomsen: So do you want to start at the beginning of geothermal?

Craig Shields: That would be great, if you don't mind.

PT: Sure. The first geothermal project was about 1905 in Italy. A farmer drilled a well, hot water came out, which turned to steam and the concept of putting a steam turbine on that to produce electricity was created. Soon we had flash technology, where water comes out of the ground, turns to steam and you put a turbine on it. And there are projects like that in Northern California at the geysers; there are projects like that in Iceland, in Africa—but they tend to be unique anomalies.

Ormat Technologies became a company in 1965. Our chairman decided that there were probably more stable ways to produce electricity and started to work on a heat exchanger and a turbine design utilizing what's called the organic Rankine cycle. The cycle simply creates a secondary loop; where there are deviations in temperature, you can heat a working fluid which does the vaporizing, which builds pressure and turns a turbine. He first implemented this on a solar project in Mali, Africa. It was technically a success, but commercially not that attractive, so he turned towards geothermal.

So this is the idea of having two temperatures heat a secondary working fluid—you can use hot water from the earth or you can use hot air from an exhaust stack or you can use any kind of difference in heat to produce electricity. But we start to utilize this organic Rankine cycle in geothermal power plants. And it created a rebirth to the industry, because you don't necessarily have to have water so hot that it turns to steam. You can use hot water to heat the secondary working fluid, you can then re-inject that water without letting it flash and go into the atmosphere, back into the reservoir, heat it back up and reuse it indefinitely. This working fluid that you heated it will vaporize for you instead of the water, and it will turn the turbine. When it comes down the back side of the turbine, you can capture it, re-condense it and reuse it.

This binary technology allows for the development of much more moderate temperature geothermal resources. And so, projects are born in the United States that were previously non-financible. So, skipping ahead a little while, we start proving up these resources and in the last decade Ormat has been responsible for over 75% of the new geothermal installed capacity in this country.

CS: I get it now. Thanks. So let's talk about the technology now for just a second in terms of its current installation. So the concept is still, you're using two separate tanks of fluid, one of which is getting heated....

PT: Yes, by the brine. So you have a real simple tube and shell heat exchanger. The hot water is around the outside of the working fluids in these tubes; it gets warm, it vaporizes at a lower temperature, about 270 degrees Fahrenheit and that is what goes into the turbine to make the turbine spin. The hot water, with all of its special characteristics, being whatever it may be, is never brought out of pressure; it's never introduced in the atmosphere—and then it's re-injected back into the reservoir. This makes these moderate temperature resources much more sustainable, because you don't have evaporation; you're not losing a lot of water from the reservoirs.

We've been developing projects for the last ten years doing that. Ormat owns and operates about 500 megawatts in the United States; we're responsible for about a gigawatt of generation around the world. And keep in mind that this was during a time period where there wasn't a really credible commitment to renewable technologies—for the last 8 years. Now we're seeing a new commitment to renewables, including geothermal.

Now here comes a paradigm shift—a technology that can greatly alter the course of geothermal development going forward. Dr. Jefferson Tester at Massachusetts Institute of Technology came up with this idea that you could start to engineer geothermal resource reservoirs. We need three things to make what we call hydrothermal reservoir work, we need heat, water, and permeability. Well if one of those three ingredients is missing the idea of the MIT report is that you can engineer a solution. If there's no water what stops you from pumping water into the reservoir? If there's no permeability what stops you from fracturing the rocks to create better permeability?

If you can do that, you can produce a hundred thousand megawatts from geothermal technologies in this country because if you go two kilometers below the Earth's crust you find heat almost anywhere. This is really looking at raw potential. DOE looks like they are trying to explore that with new subsurface drilling techniques; a lot of this new DOE money was looking at the first steps of looking at these engineered geothermal systems or what they call EGS.

CS: Well as a friend of mine says, "The issue isn't 'is there plenty of clean energy.' The issue is 'what are you willing to pay for it?'" Does this compete well against other forms of energy?

PT: I think that's a very good point. And you know, simply put, geothermal is a nice technology to talk about because it's relatively simple. We bring up heat from the Earth, we need a medium to bring that heat to the surface, and instead of having a natural gas fired Bunsen burner to create steam, we're using the heat from the Earth.

Simply put, the farther we move eastward, it's kind of a reverse manifest destiny, we started in the western United States where we had lots of faults and lots of geologic activity, pushing up on the Pacific and North American plates. As we move eastward in the United States away from the Ring of Fire, we have to drill deeper, which is costlier. And we're finding lower temperature resources. So the question is where is the breaking point on these cost-benefit analyses.

Then you run into a whole bunch of other factors that have impact. The sudden drop in natural gas prices has hurt geothermal developers. A year ago it was much more attractive. On the other hand the federal government discussing a renewable energy standard and a carbon tax that puts a price on the intrinsic attributes of fossil fuel generation starts to make geothermal look a little more attractive. So you balance these things as you look toward developing these resources.

Geothermal as a renewable is one of the most attractive resources to develop if you have the resources. Once you find good heat, good permeability, and lots of water to bring that heat to the surface, you have created a base-load power plant—a power plant that produces power 24 hrs a day, 7 days a week, 52 weeks a year. You can typically get a contract with a utility to purchase that power at a reasonable rate with escalation over a fixed time period. Utilities, at least the ones we're familiar with, have been so bold as to say when they look at geothermal power plants they don't look at them any differently than their fossil fuel counterparts.

CS: The only reason they would look at them differently is with respect to just reliability, right? Everybody wants green, but it's best if those power sources aren't variable and unmanageable.

PT: Well, they do want those; they just serve a different purpose. But when they look at our integrated resource plan, we are actually more attractive even than their fossil fuel brethren; our binary geothermal power plants operate on average at a 95% capacity factor.

Where you have this resource these power plants have proven themselves over a 20-year period to be one of the most reliable power sources out there. Going back to the engineering side, the power plants themselves are very low entropy and again, decoupling the turbine from the brine has just added longevity to these projects. You know if you think about a geothermal power plant, the highest temperatures are around 300 to 400 degrees Fahrenheit, and the highest pressures are 150 average PSI. They don't have a lot of things that can go wrong with them. But when you're dealing with a huge boiler in a massive fossil fuel plant, there's a lot of O&M (operations and maintenance) that goes into producing that the amount of power.

Here, there's near zero, unless we're doing maintenance on the plant. We have what we call fugitive emissions. Because the cistern is pressured, when we do our yearly maintenance and we disconnect...it's like a propane tank to your barbecue. When you unhook it you get that little "psst" that's about what we have to deal with.

CS: So in other words the temperature coming out of the earth is only 400 degrees Fahrenheit?

PT: Our place in Reno is actually a bit lower than that. They tend to sit around 300 degrees Fahrenheit. We run that hot water through our plant and we re-inject 100% of the geothermal brine back into the earth at about 200 F. So we are extracting about 100 degrees Fahrenheit from that water.

People will say, "Why don't you cool that water down to ambient air temperature 60 degrees Fahrenheit?" There are a couple engineering reasons and the first is, we don't want the particulate matter that's in that deep brine to come out of solution. The more we cool it off the more the heavy metals and things want to come out of solution and cause scaling problems for us.

The second is that the water is like the medium for the heat. All the heat is in the Earth's crust but if we re-inject freezing cold water into the reservoir it will take longer for it to heat up. So we need to find a balance between our re-injection and our production. If we run the water through the system too fast it doesn't have time to fully heat up and if we leave it down there too long we're not extracting the maximum amount of energy. So we have a lot of reservoir engineers who are monitoring how long it takes from the time of injection to the time of production and how many millions of gallons of water we can pump through our system before we start to see any detrimental affects on our power plant.

This brings me to the most important part. The huge differential between fossil fuel plants and renewable power—at least geothermal—is we have a symbiotic relationship with the reservoir. If we don't maintain it and protect it—if we start to cool it off, we will see an impact on our power plants and produce less electricity and make less money. So from our perspective, maintaining that reservoir is our number one priority. If we were to yank our power plants off of it today we would hope to re-inject 100% of that water back to the reservoir and let it just start cooking like it had done for thousands of years.

CS: That "cooking" is largely due to the pressure right? In other words as the earth was formed the gravitational attraction of one particle to another just heated the whole thing up. Is that right?

PT: I'm not a geologist. So I think when you're talking global, you know, center of the earth heat that you're about right. That's what gets you up to the 7000 degrees or 9000 degrees Fahrenheit or something like that. In the crust, I think it's simpler than that—it's where the magma and the formations of our earth's crust are pushing together. They're bringing continental crust type heat closer to the surface. When you look at where mountains are pushing up and so forth there is thermal activity underneath there creating that. We're simply trying to tap those places that are closest to the earth's surface.

CS: What is the typical depth?

PT: The depth is between a thousand and eight thousand feet deep. So we're not penetrating anything below the earth's crust. If we could drill you know, 20 or 40 thousand feet it would be a different story. But geothermal wells are very big because of the volume of water we have to move, which really inhibits the depth at which we can drill economically.

CS: Tell me a little bit about that, please. I understand there've been some breakthroughs in drilling recently. What are they? How would you explain them?

PT: Well I don't think there have been any big breakthroughs recently. I think the geothermal industry today is kind of where the oil and gas industry was in the 30's. You know, where the oil was bubbling up on the surface. We have developed most of the resources that had some kind of surface manifestation.

CS: Where you could actually see it?

PT: Yes, where you can see it. Where the snow melts first or where there's a big hot spring sticking out of the ground. We are moving into the territory of trying to go deeper and develop blind resources. But the industry is still using a lot of the same techniques that the oil and gas industry used 50 years ago. And that's partially because of R&D budgets and so forth, the huge subsurface mapping, and things cost a lot of money and how water behaves at depths in the Earth's crust is a little different than oil. So, DOE just put out I think $300 million in grants to look at innovative drilling technologies, and we hope that this subsurface research will create some major breakthroughs in drilling and really reduce the risk.

You know there's two things that we need to do. One is to be able to drill deeper and two is to reduce the risks of drilling unproductive wells. If we can do those, you'll see a major breakthrough in geothermal development. The more wells we can drill, and the more successful those drills are, the more geothermal power plants you'll see.

CS: I thought I heard that one of your smaller competitors had come up with some sort of killer drilling rig.

PT: Well, drilling's not really our sweet spot, so to speak. We've only purchased drill rigs probably in the last two or three years, and hired staff to do our own drilling. We had previously contracted our drilling but competition with the oil and gas industry became so tough it was very hard to get your hands on a drill rig. The costs were becoming exorbitant. We are not, by any means, an expert driller; we do own four or five drill rigs today but there are firms out there and contractors who have been drilling in the oil and gas industry and the geothermal industry far longer than we have. Our strength really is the turbine in the power plant that we design and manufacture.

It's also project development. Because we've been doing it for so long we know what it takes to design the plant to fit the resource and implement bringing that power plant online. The only real new drilling company that I can think of is a company called Thermasource.

CS: I think that was it. How wide is the hole you're talking about?

PT: It's about 13 and a quarter inches in diameter at depth.

CS: Okay so you extract the water through one hole and you put it back through another. How far apart are those two holes?

PT: That's completely variable. They typically tend not to be right on top of each other. An analogy I use is if you had a bath tub and if you turn the hot and cold water on full blast right next to each other you get a real mixing of hot and cold. So when we model the reservoir, instead of having the cold water mix with it, we want to re-inject it at the back of the bathtub, so the cool water has time to heat up before we extract it out of the front of the bathtub. Now that can be a couple hundred feet or a couple miles; it really depends on the geology and the reservoir.

CS: I see. Is there some advanced geological way of determining what the porosity—what did you call it? Permeability? How these things are all connected thousands of feet under the ground?

PT: No. There's no magic computer program. It's a lot of drilling. We drill what we call slim holes which are much smaller than production wells; we do a lot of testing on those to see what we call "communication"—how the water flows between those slim holes. But we don't learn a lot until we're actually up and moving large volumes of water. We have circulation loss in those reservoirs and so this is part of the reason that you see what I would call an incremental growth phase in geothermal plants.

What I mean by that is, in the city of Reno, Ormat produces a hundred megawatts of geothermal electricity—enough to supply the entire residential load of the city. We developed that over a 20-year period building 30-megawatt power plants one at a time. Now people will ask, "Why didn't you build one 100 megawatt power plant?" The reason is that we think the resource might be big but if we build a 100-megawatt power plant and then find out that the resource can only sustain a 30-megawatt development, we've just lost 70 megawatts at 2 million bucks; we spent 120 million dollars on two thirds of a power plant that can't produce electricity.

And so at the same time we start moving small volumes of water through this reservoir and our reservoir engineers can start modeling and gaining information to tell us whether or not that reservoir can sustain more development than what we currently have. In the case of Steamboat, we had 30 megawatts and they said "You know what, the heat source here seems to be recharging, we have plenty of water in volume, we think we can sustain another 30 megawatts." So we built that and then we're at 60 megawatts and we say "You know what, we haven't seen a decline in heat; it seems to be handling the volume just fine; we're confident we can do another 30 megawatts," so we built that. Now today at 100 megawatts, our engineers are saying, "You know what, we're not seeing as much recharge and we're not as confident that this could handle another 30 megawatts," and so instead of taking a risk, we back off.

CS: Okay, well this is wonderful stuff. So your sweet spot is your turbine technology. It strikes me though that this must be, if not identical, then very close to solar thermal turbine technology, right? You're taking a hot fluid and turning it into electricity?

PT: A solar thermal can use one of two turbines: a steam turbine or Ormat's binary system. We provided the turbine for the first solar thermal project in more than 20 years in Arizona for the Arizona Public Service. They heat a thermal oil, they run it into our heat exchanger and our turbine sits there and produces one megawatt test project. Now the big solar project in southern Nevada, and the ones that they're proposing to go to the hundreds of megawatts or gigawatt phase, they're looking to use steam. And put in huge steam turbines and so they've really have to get that heat way up there.

CS: So, where do you see this going?

PT: People are waiting, hoping for a dot.com boom if you will—for California especially. That phrase kind of rings true. When you talk about the dot-com boom, you had people who wanted to communicate and share information. We'll call sharing of information a resource, and there was tons of resource out there. There were millions and millions of bytes of information that wanted to be shared but there wasn't a real good technology to advance that. You develop the computer and the Internet and all of a sudden you've tapped this unbelievably huge resource and you could sell a million computers and the dot-com boom takes off. I think my impression, and this is my personal opinion on geothermal especially, is that it's a bit reversed. The technology to develop geothermal resources or exploit the resource has been here for a hundred years.

A steam turbine spinning the steam or binary turbine to produce electricity, we know how it works, it's field-proven, there are millions of run-hours on it, so why isn't it taking off? Because the issue we have is finding that resource—that plethora of resource all over the country. We're a little different because I think solar is a little more like a dot-com boom. You can say there's tons of solar radiation out there, it's quantifiable, we know that if we could cover the earth's surface in solar panels we could power the United States. But if the technology could become cheaper and more efficient, you could probably sell a billion solar panels.

For geothermal, we have to find that resource, so for us it's drilling deeper, tapping these hot pockets of water or engineering them, and bringing them to the surface. Because once we do that we're pretty confident that the technology would have more than adequate resources. What defines geothermal development is the development of the resource. Drilling deep into the earth and finding water pockets that have good permeability and good heat and are closed systems so we don't have huge leakage is inherently difficult and costly and hard to do. That won't change anytime soon unless there's some paradigm shift in drilling and looking deep into the earth and finding these fractures and drilling into them; that will always be the difficult part. It's not so much the technology.

Now if you want to get futuristic, there might be things that we can't see 40 years from now. Maybe you run fiber optic cables into the heat and transfer the heat with no loss and do funky things. Maybe you pump super-critical CO_2 into the reservoirs and use that to transfer the heat and things like that. But for us it's not so much improving the technology as it's developing that resource. That is a huge hurdle. And a shot in the arm is that DOE has put money into this, saying that Ormat will compete with GE and Fuji and anybody else to build a better turbine. Market forces will demand that we build better equipment.

What hasn't happened in the last decade or so is that crucial subsurface research. The industry spends millions of dollars in exploring dry holes, and small geothermal companies can't deal with that.

CS: Right. I'm actually surprised to hear that you don't have sonar or whatever they use in oil exploration—where they can look for big pockets of stuff that isn't rock.

PT: Yeah, big voids. And we do the same, we look for voids, but there's not always hot water in those voids. Sometimes they can be natural gas or oil. And with oil and gas exploration, to completely oversimplify it, you tend to find those extra resources in older geological structures that tend to be in sand and more friendly drilling environments. We tend to be drilling in the bedrock that is pushing up and creating mountains and very hot, unfriendly resources.

So when a drill rig goes to drill a natural gas well in Texas it's drilling through sand and old sedimentary rock and things and can penetrate very quickly and very deeply. And it's also a much smaller-bore hole. We kick it up a notch and say how can we make our lives miserable? We drill a bloody big hole in very hard, unfriendly bedrock and we're going into the hottest hellish places to try to find these resources. So we chew up drill bits, we pay tons of money to be out there because it takes three times as long and costs three times more money—and then we might not find as much water and as much permeability as we expected.

We have many wells that are very hot at the bottom but there's no water. Or they're very hot and there's lots of water but we can't move the water through the reservoir as we expected. And then we're just out two to seven million dollars per well. So the EGS technology might come along but it needs to come along in conjunction with the typical hydrothermal drilling. It really will reduce our risk; we could drill twice as deep if we could get half as many dry wells.

CS: How interesting. Let me change course a bit here. Do you believe that the development of renewables generally exists on a level playing field with big oil, gas, and nuclear? Is the political process as it applies to you essentially fair?

PT: Americans believe that they should get as much electricity as they want, for as cheap as possible—it's a birthright. What if we had made a commitment to renewable technologies during the first oil crisis in the 70's? We'd have cheap baseload geothermal. But we didn't. It's because politicians don't look 20 years down the road; they look at the next election cycle.

That's the tricky part. They need to get re-elected in two years and energy, and high energy prices today don't necessarily equate to re-election in two years.

Renewable technologies are missing a lot of the negative attributes of fossil fuel emissions. But we didn't put a value on those before; in the economic market they were intangibles that had no value. People wanted megawatts and they wanted them cheaply. Now if you take a fossil fuel plant today and say "Your emissions will cost you money," you have done a great service to those technologies that don't produce those emissions. Whether you care about global warming or asthma, there should be an incentive to drive those emissions down. But it's very hard to tell politicians to look 40 years down the road when they'll be long gone.

CS: Yes I can understand that. Well that is a very good, measured, and thoughtful answer to a difficult question. And I really do appreciate it Paul.

PT: Fantastic, Craig. Thanks for the opportunity.

For more information on this contributor, please visit: http://2greenenergy.com/renewable-energy-facts-fantasies/.

HYDROKINETICS

An extremely senior engineer referred University of Washington's Dr. Brian Polagye to me as "perhaps the world's most senior researcher in hydrokinetics"—and I must say that he did not disappoint. His work focuses on responsibly harnessing the kinetic energy in moving water, in particular, developing a better understanding of the practically recoverable resource for tidal streams.

Craig Shields: Dr. Polagye, maybe I could ask you to take me through the kind of technological breakthroughs that have been occurring here—you know, the most important issues with respect to physics and engineering that will ultimately enable hydrokinetics to play an important role in the ensemble of renewables. But first, is there enough potential energy to make a difference? Let's say within the continental US—is that the totality of the potential energy– the weight of the water times the height from which it is falling—does that get us anywhere near the 5.4 terawatt hours that we consume here in the US every year?

Brian Polagye: I don't think there's any one energy solution that gets you all the way there. I mean, you wouldn't legitimately expect to replace all the power we currently consume with a single source like in-stream river hydrokinetics. That being said, I think that river, tidal, wave, ocean current, all of these can make a valuable contribution, either nationally or regionally, to the electric grid. So I think it's important not to discard an idea simply because it doesn't solve all of our problems.

CS: OK, but at the same time, I would submit that even if you got every joule of energy out of hydro, if you're still only at a couple percent, there's reason not to be as excited about it as you would if you were looking at, for instance, solar-thermal and saying, "Well look, if we get 1/6000th of the energy bestowed by the sun every day, we're home."

BP: That's true. If we could do that, that would be fantastic. If you had a way to integrate it into the grid and deal with the storage problems and the fact that the sun only shines for at most 50% of the day. If you overcame all of those problems, that would be absolutely fantastic. I view river hydrokinetics and tidal hydrokinetics—and less so wave—because people are still trying to work out how to take the power out of waves effectively—but for flowing water applications, tidal or river, we have a pretty good idea how the technology works. We have a pretty good idea how to extract power. And it's a smaller total market than say solar-thermal or solar photovoltaics, but there's less R&D required to actually get that power on the grid.

CS: Right. Moreover there's 24 hour a day service.

BP: Well, for tidal it comes and it goes, and for rivers it varies with the season. There's virtually no renewable energy—with the possible exception of biomass if you're considering stockpiling wood—that can really provide 100% base-load on its own without some sort of storage technology. And wood basically has its own storage technologies.

CS: OK. Well take us through, if you don't mind, what is the current thinking with respect to river and tidal. How do we extract power? I mean, basically you're talking about Faraday's discovery 180 years ago about a moving conductor and a magnetic field, but I'm sure there's more to it than that. What is the cutting-edge in terms of extracting power?

BP: Well, I can mostly talk about tidal; that's where my expertise and research is. I'd say the real cutting edge technologies on tidal are not necessarily in the turbines themselves, as far as extracting power. We know how to extract power from moving water—or from moving air. There are various modifications. You have different blade designs, variable pitch versus fixed pitch, the use of ducted turbines, vertical axis versus horizontal axis, and all these variants. But I think ultimately the kind of technical advances that we're seeing are advances related to the rapid deployment of this technology, environmental mitigation measures, and, I would say, the survivability and reliability in the marine environment. I think those are where the big question marks are and where people are making real strides.

CS: Interesting. Can you give me an example so I can envision this? I could understand why, for instance, survivability and reliability in a marine environment would be something of a trick. How do you address that?

BP: Yes, well, and a very important trick to get right. Because the cost to go out and do maintenance in a marine environment is so much higher. I'd say a reasonably good example, and this is more of a kind of a "compare and contrast": Open Hydro Technologies, which is a developer based out of Ireland, recently deployed a turbine in the Bay of Fundy. And this company has worked up its own proprietary deployment barge and installation strategy that allows them to put a turbine on a gravity foundation on the seabed over a single tidal cycle. This is a much faster installation time than for say, a foundation that requires you to drill or drive a pile into the seabed. So I think that's kind of an interesting advance in actually getting these devices in the water. Rather than it being a month long procedure, they're getting it done in only a few hours.

You do it during a period of weak tidal currents. You're operating on a 15-day cycle where the tides vary between what are called neap and spring tides. Every day you get your two high and low tides, so two ebbs and flows, and then the strength of those ebbs and flows will vary in that 15-day cycle.

Ideally you want to do installation during a period of neap tides, when the tides are at their weakest. Generally you're not able to get the installation done during slack water. At many of these tidal energy sites, slack water is something that doesn't really exist or only lasts for only about five minutes; you'll never get a device deployed that quickly.

That being said, the design that Open Hydro is using has certain limitations. You can only have a turbine so high in the water column on a gravity base. It's easier with a pile foundation to have the turbine closer to the surface where the power density is higher. So there are lots of different engineering trade-offs, but I'm really excited about a lot of the new innovations that I see with respect to putting turbines down and keeping them in service.

CS: And you were talking about environmental mitigation. So in other words, you obviously don't want it to chop up fish, and so forth.

BP: Chopping up fish is one of those things that you immediately think of when people are concerned about giant turbines. They see blades turning in water, they worry about the speed at which the blades turn and their potential for strike. Ultimately, strike is probably not the biggest concern for a lot of these projects. If you look at how the flow field behaves around the turbine, as water approaches the turbine it flows down and you actually end up with your highest velocities being diverted around the turbine. Strike is potentially less important than things like the acoustic footprint of an operating device. Underwater sound travels very far and many marine mammals have very sensitive hearing.

The other general types of environmental effects are avoidance, which is generally triggered by noise or the device wake. Then there is aggregation, where if you had marine life colonizing the surface of a device structure, say you have a pile down on the seabed or a gravity foundation and it gets colonized by marine life and that ends up attracting things into the area. That can also especially be a problem if it's attracting species away from their natural habitat into artificial habitat or concentrating fish and drawing predators.

CS: Right. I understand there's an issue with giving predators a place to hide.

BP: Yes, potentially giving predators a place to hide or concentrating predators in areas where they aren't naturally. It's just looking at ways in which these devices operating would affect the natural environment. Until there are more devices in the water, you can conjecture endlessly about what the problems might be but you have a very hard time of actually reaching any consensus on really when things are a matter of concern and when they're not.

CS: Oh, I'm sure of that. If there's debate about global warming, Lord knows there's going to be debate about something like that.

Isn't tidal restricted to pretty much northern and southern latitudes? You get higher tides in the Puget Sound than you do in the Panama Canal, right?

BP: Yes, you're looking at the extremes. New Zealand has fantastic tidal resource as well. Australia has some. China has some, actually. And then you're talking about the UK, Canada, and North America.

CS: Well isn't that somewhat problematic in and of itself, insofar as most of the population centers aren't at those latitudes?

BP: Well, there are certainly areas where that exists, the Aleutian Islands being a fantastic example. Great tidal resource, but few places to use the energy. But there are other places like the Puget Sound; there's a bay in New Zealand which sits right on top of a pretty large load, and the UK is obviously a good example. So it's a challenge, but it's not an insurmountable one. I mean, you could look at solar and say "Well, the best solar resource is in the Sahara Desert," but that doesn't mean you would never develop that; it just says you need to generate the power there and transmit it somewhere else.

In some cases, like the Aleutian Islands, it'll probably never be economic to generate power there and transmit it back to anyone who can really use it on a large scale. But in other cases it is. If renewables were easy we'd be doing them already, right?

CS: Exactly right. Let me ask you about run-of-river. It sounds like you would have the same concatenation of environmental challenges. What are the major differences as you see them?

BP: Well, obviously, freshwater versus saltwater environments. The ecology is completely different. The type of materials you can use can also vary considerably. Certain metals that are corroded rapidly in saltwater are not in fresh. The biggest difference between river and tidal is river has more seasonal variability. And it's unidirectional flow. Those are big differences. It's similar technology, it's just a different environment to be deploying in. Shallower water depths is another big difference between tidal and river.

CS: I would think unidirectionality would be a plus.

BP: It certainly is, it's good that it's unidirectional but the flows tend to be lower velocity.

CS: Oh. I would also think that seasonality would apply to smaller rivers that wouldn't be good candidates for this in the first place.

BP: Not necessarily. I mean, with all river flow, unless the river is dammed and there's a large reservoir that's regulating flow upstream, there's definitely seasonal spikes associated with most rivers. When it rains, that's when the water comes out.

CS: I guess I'm used to looking at the Ohio River or something and thinking it's pretty much 24/7.

BP: Well, the river is flowing, but I'm sure if you were to look at the stream flow over the year that you would see significant variation. Now something like, say, a river that is heavily dammed in, you know, say somewhere in Colorado, maybe in that case if the reservoir is basically metering out a constant amount of water throughout the year and you're downstream of that, then it is effectively constant.

CS: Where is your work concentrated now? What gets you up in the morning?

BP: About thirty different things at once. I'm a faculty member with Northwest National Marine Renewable Energy Center. It's a DoE funded partnership between Oregon State University and the University of Washington. We look after tidal issues—broken down into four areas. The first of which is trying to optimize devices and arrays of devices, the second of which is trying to characterize tidal energy sites and devices as they operate in the field, the third is looking after environmental considerations, and the fourth is looking after reliability and survivability of devices. Most of our work in that final area has been focused on looking at corrosion and biofouling for different materials. So the work that I'm involved in is the site characterization piece and the environmental effects piece. And I have a number of different things going on in each of those areas right now.

CS: I would guess a lot of this gets into material science.

BP: Oh, some of it does, sure. Though, if you think about wind turbines, those use relatively simple materials. You know, the internal frame for some wind turbine blades is actually Balsa wood. So, it's not necessarily all really high tech materials.

CS: In the research I did for a white paper on hydrokinetics a few months ago, I came across a horror story where a company called Verdant struggled through years of permitting and ended up having to pay back Consolidated Edison for the power that ConEd wasn't able to sell. What do we have here? Over-regulation? Corruption?

BP: No, I don't think it's corruption. I think it's just that the processes that exist today are not well suited for pilot projects like what Verdant is trying to do.

This is really a tough question. And I definitely have sympathy for the regulatory agencies. The saying is that you're not doing your job well as a regulatory agency if you're not being yelled at by everyone. The regulatory agencies in the US are charged with permitting some of these projects. On the one hand they're under pressure by device developers for putting up so many hurdles to getting devices in the water, and on the other hand they're under pressure by environmental groups for allowing any devices in the water. So, depending on whom you talk to, regulatory agencies are either being wildly permissive or incredibly hard-line in their approach to these projects.

CS: Ok. I guess it was the six years that it took Verdant to get their permit. They have funding sitting around, they've got potential solutions, and they've got six years before they can drop a device in the East River. My point is that six years is an eternity—the whole world is a different place six years later.

BP: Yes. I don't have the answers to this, obviously, because it's a tricky issue. But imagine, let's say for example that regulatory agencies were to allow a large array of turbines to be placed in Puget Sound without any studies or monitoring and say "Well, we don't really know how these work, but we think there's probably not going to be any significant effects, so let's put it in the water." And an Orca washes up on a beach. Maybe it died from natural causes, maybe it was struck by a ship, or maybe it was struck by a turbine blade; no one knows. That could either set the industry back by years or kill it entirely.

I think as these projects go forward and we learn more about the environmental risks, the permitting timeline will eventually compress. But the initial developers who went out there without any sort of government support and tried to do it on their own, they got a lousy deal.

Before the federal government got involved in funding some of the environmental studies and supporting these developers, people trying to run companies on venture capital funds or, in the case of Verdant, mortgaging their own houses, that's a tough situation to be in. I think the DOE's continuing support is essential for keeping the industry on track. It's definitely helpful to have DOE pushing behind some of these technologies and in cases where a regulatory agency says we really need to understand something better and a single device developer can't handle the financial burden of that, it's nice to have support. Regulatory agencies can ask for all manner of studies, some of them appropriate, some of them not. But in the cases where a study is appropriate, but the cost is high, it's important that the DOE is able to get behind these studies and help to make them happen to move the entire industry forward. And then the lessons learned from DOE funded studies end up being broadly disseminated; they don't just stay as a lesson learned with a particular developer. So I think DOE has an absolutely critical role that it plays here.

CS: OK. Well, staying on politics for just a second before we get back to science and business, what have been the major differences between the Bush versus Obama administrations?

BP: I think there has been a very long-term shift towards additional support for renewables. It was already happening at the end of the Bush presidency. I haven't seen a real digital change. It certainly has continued to increase, but I feel like it was already on that trajectory.

Ultimately having somebody at the top say, "We want this" helps a lot, but when it comes down to individual projects, a lot of the people who have been in these agencies and organizations for several different presidencies are the ones who ultimately make the calls on where resources get allocated. So the people from the top certainly help to set the tone, but if the people below them think it's a bad idea, they'll still say it's a bad idea.

Having said all this, hydrokinetics is such a new technology that it's the sort of thing that when people try to pick a real keynote issue to really focus on for their presidency or for their lead in an organization, they're probably not picking tidal hydrokinetics. Over time that may change, but for now it remains just one of many developing renewable energies.

CS: Do you regret that we have subsidies for things like corn ethanol? Do you ever wonder where we would be if we had used those subsidies for hydro, or pick one..solar-thermal or ..

BP: Or PV or advanced wind or offshore wind. I mean, that's the problem with renewable energy. There's so much. You section it up into all these pieces and then you section it up further and further until you get to a really stunningly vast array of ways to generate power. And this is not a problem that just the US is grappling with, but all world governments. If you're going try to fund the development of this technology, do you pick a technology out of the thousands that are available and say "We are committed to this one type of renewable energy?" Or do you basically try to foster as many possibilities and see what the free market decides makes the most sense?

CS: Yes, and the latter obviously is the weapon of choice. But personally, if I were king of the world, I would be closer to the former than the latter. I know that sounds radical, and I've never heard anybody else say that, but it just seems like we're marching in so many different directions at the same time...

BP: That nothing ends up getting done. No, it's a legitimate concern, but it's the way everyone has decided is the fairest way to try to do this.

CS: Yes. Well, what, if you were a betting man, what would you say the renewable energy world is going to look like ten years hence?

BP: Boy, that's an interesting question. I think it really depends on what happens in the next few years. In the US, if we figure out how to effectively permit projects, and they go in the water, and there are very minor or mitigatable environmental effects, I can see hydrokinetics playing a substantial role, yet probably more on a regional scale than a national scale. If projects do go in the water and there are problems, or DOE decides that they are not as interested in continuing to fund the development both of the technology and the environmental studies, I can see the technology withering and dying in the US, but continuing to be developed in other areas of the world. So I think the next few years are really critical and it's really important to make sure that the projects that are going in the water over the next few years are done right, they're instrumented well, and that we learn a lot from them. That's the single best thing that we could do is to learn as much as we can from the few projects that are going in the water.

CS: Do you mind telling me a bit about wave energy? Is the fundamental technology now the creation of pressure in a cylinder that is then pumped on shore and used to turn an on-shore turbine?

BP: That's what I would call a near-shore or on-shore wave energy device. There are also people looking at off-shore wave energy. The general idea with wave energy is that over very long stretches of the ocean, the wind acting on the surface creates chop and the ultimately swells of variable height and period. And that up and down motion is what you're harnessing with wave energy converters. Then you need ways to turn that up-and-down motion into electricity, the same way with tidal current—you're looking at ways to change that horizontal motion of the water into energy, in this case you're looking at ways to change this relatively rapid up-and-down motion into energy.

CS: OK, but what is the best technology to do that now? If we had to do that next year, what approach would we take?

BP: We would scratch our heads and try to choose from the twenty different technologies available. The truth is that no one's figured out what the best technology is for wave. One of the reasons for that is, you look at say, tidal current, we can look to what worked in the wind industry and what didn't and we can really get a leg up on which technologies are going to work best. For wave, it's totally different in terms of extracting energy. We've been experimenting with wave seriously since the 1970's, but as far as a technology that we're ready to converge on, we're not there yet.

All of these ideas have different benefits and disadvantages. If you took all the technology that we have today, all the different approaches people are doing for wave, there's probably a piece of each that has it exactly right. Over the next ten to fifteen years, you'll see a general convergence of devices and where they're deployed and the type of device and the type of power take-off and the type of mooring. But right now we're very much at the stage of a lot of different people trying a lot of different things and some of them are going to work and some of them aren't.

CS: From my interview with NREL, they have me convinced that there is a wonderful openness and camaraderie between the public and private sector and of course by extension, I suppose, the world of academia. In other words, that there's very little parochialism in terms of getting this done, which I find almost hard to believe. I'm thrilled to contemplate this kind of harmonious relationship—maybe I'm too cynical, but it's almost hard to imagine that it's true.

BP: You know, there's a lot of really good interplay between the public and private sector in this, and there are other areas where the interplay could be better than it is—which I think is probably true of any venture. For example, many of the universities in Europe have developed their own wave or tidal energy devices. And this is because they fundamentally want to understand how these devices work. They want their students to research these devices, they want their students to do research that's publishable and they can be very open and transparent about it. Device developers, on the other hand, would love to see the research done, but don't want to see all the results published because it gives away their competitive edge.

So the device developer will say "Well sure, we'd love to have you work on this project, but you'll have to sign a non-disclosure agreement and your students can't publish." Well, that doesn't work for the university, so you end up with a situation of device developers that could certainly benefit from some of the research being done in academia but aren't willing to be as open as academics. There are academics who want to be completely open about things which effectively destroy the trade secrets that are the lifeblood of these companies. There are definitely cases where it works well and there are other cases where you wish it would work better, but it's a really tricky issue.

Generally the situations that I've seen where it works well are public-private partnerships, where everyone's worked together for a number of years and has developed a level of trust that they don't need necessarily a legally binding agreement to do research anymore. They will say, "I know you, I trust you, and we're all in this together." When that works, it works fantastically well. But it just takes a while to establish those relationships and in an industry that's barely five years old, there's not a lot of those around. So everyone's very congenial and we all work together, but there are certain things you know not to ask for or that you know are probably non-starters for a research project because of those intellectual property issues.

CS: Is there a great deal of data available on hydro, along the lines of what 3 Tier has done in wind?

BP: 3 Tier's expertise is largely, as I understand it, in resource forecasting—in creating high-resolution wind maps and solar maps. I'd say the work we're doing now, the measurements we're doing, are feeding into a very similar database, but it's a very nascent effort. Right now the hydrokinetic industry is probably not large enough for Three Tier to really say "This is a great place for us to jump into." As it gets larger, I have no doubt that Three Tier will start to look to see how they can take their existing expertise and apply it to new industries. They have a very good group. I have a lot of respect for them.

CS: Well, this has been fantastic, Brian. I can't thank you enough. Is there anything else you'd like to tell me about this that you think readers would be interested in hearing?

BP: I can think of lots of things, but you know I think the important thing to realize is that the regulatory environment is a tough nut to crack. Every country that is developing these technologies is eventually going to have to face it. In the US, I think we're kind of hitting it at an earlier stage, but eventually in Canada and the UK, as people try to deploy larger and larger arrays, they're going to run into similar questions. And I think that in some ways the work we're doing in the US is actually catching up and getting ahead of what the Europeans know about the environmental effects of these technologies.

CS: Great stuff. Obviously, it's important that the US plays a leading role here in renewables. Ray Lane, the great venture capitalist at Kleiner Perkins told a big audience at the "Business of Plugging In" conference in Detroit last fall that what we've done in the internet with Google and Microsoft and Oracle and so forth is good, but if we don't do something similar here, we'll be in bad shape—and that we have to stop talking about it and start doing it.

Thanks very much and take care of yourself.

BP: OK, you're welcome Craig. Talk to you again soon.

For more information on this contributor, please visit: http://2greenenergy.com/renewable-energy-facts-fantasies/.

COLD FUSION

Here's an interview with Wally Rippel, known to renewable energy enthusiasts everywhere as one the best informed and most committed proponents of electric vehicles and environmental stewardship more generally. In the course of an earlier conversation I had with Wally on electric transportation, he happened to mention cold fusion, and it occurred to me that readers may wish to learn more about this latter subject.

Craig Shields: When I was out at your place a couple of weeks ago, you mentioned that you believed that cold fusion was actually legitimate science. I know this is a controversial topic, but I was hoping that you wouldn't mind letting me understand it from your perspective. May I ask you a few quick questions?

Wally Rippel: Absolutely. Let me begin by asking you, Craig: How much do you know about this?

CS: Oh, I can tell you everything I know in just a few sentences—and even that may contain errors. Here goes. If you electrolyze heavy water, i.e., water made of deuterium and oxygen, in the presence of a palladium anode, you wind up with more energy that you would expect—and more than you could explain with standard chemistry—and you get products that suggest that a nuclear reaction took place.

WR: Yes, that's right. Cold fusion is a concept that began in 1989 with the experiments of Fleischmann and Pons at the University of Utah. The world had wrongly assumed that fusion could only happen in the way it was developed in the hydrogen bomb, i.e., 100+ million degrees Kelvin and extremely short time periods (10^{-22} seconds); it was simply an intellectual error to assume that anything else could not be fusion. Nuclear reactions provide 10^7 times the energy found in chemical reactions (10 million eV versus 1 eV). So, in truth, if there are any large amounts of energy and any nuclear products, e.g., gamma rays and neutrons, it can only be a nuclear reaction.

Both in the conventional "high temperature" and the "cold fusion" environments, two reactions are noted:

$$D + D \longrightarrow He_3 + N + photon\ (gamma),\ and$$

$$D + D \longrightarrow He_4 + photon\ (gamma)$$

CS: Why doesn't it just make He_4 every time?

WR: It's because of quantum mechanical effects, i.e., the Schroedinger Wave Equation—combined with the conservation of angular momentum that make certain outcomes far more probable.

In the case of conventional fusion, the preferred reaction is #1 (99.9999% of the time). But reaction #2) does take place roughly one event in 10^6.

In the case of cold fusion, the preferred reaction is #2 (possibly greater than 99.9% of the time). These alternatives are referred as "branches" and the term "branching ratio" is used to identify the relative frequency between the two reactions.

CS: OK, but where does the energy come from? To me, it looks like that in either case, both sides have the same number of protons and neutrons. Doesn't the energy come from the mass-energy equivalence?

WR: Yes. There are no whole particles that annihilate, but the mass of components on the right side of each reaction is about 1% less than that of the left side. You know that a charged battery weighs more than a discharged battery. It's the same thing here. The difference is the energy that is stored in the electric fields. The energy is the integral of that electric field.

CS: OK, thanks. I think I understand the physics now—at some level, at least. But why is this so controversial? Why do people believe that this is bad science, a hoax? I would think that it either happens or it doesn't.

WR: Actually, the human side of the equation is even more interesting than the science. There has been a calculated effort to discredit the idea of cold fusion. For obvious reasons, cold fusion threatens existing energy-related interests, and those interests have been intensely aggressive with throwing people off the trail.

CS: But I heard that the people who looked into this couldn't find nuclear products.

WR: Ah. That's simply not true. The researchers from Cal Tech and MIT did find nuclear products; they fudged the numbers to get the DoE off the case. The US Navy and Lawrence Livermore have also found clear evidence of nuclear results. And even when they admit they find nuclear products, they would say things like, "Oh the engineering on this will be really hard." I'm not saying it will be easy, but that's like saying, "There is a trillion dollars in that safe over there, but there is no use trying to get at it because the combination lock might be hard to open." All these ridiculous ideas are a result of the enormous pressure to agree with the idea that cold fusion is a hoax.

CS: To me, this sounds a lot like what I hear about the science—and the politics—behind global warming. I hear that even extremely senior people are ostracized all the time for not conforming to the mainstream viewpoints on the subject.

WR: Exactly right. Dr. Peter Hagelstein at MIT, best known for his X-ray laser, is also a strong proponent of cold fusion. He's been isolated from the entire scientific community because of that belief.

Some of the people who had investments in cold fusion testified against it, apparently so that they could maintain majority control of the development of the technology.

Yet despite all the active attempts to divert attention away from cold fusion, the technology carries with it an enormous amount of credibility—but most of it is very quiet. For instance, the US Navy has performed experiments producing neutrons in groups of three. Also, there is an internal memo within DARPA in which they clearly state that they believe that cold fusion is real.

Energy companies have worked hard to discredit cold fusion, though, with the Obama administration, they will have a harder time doing this than they did under Bush.

Cal Tech has done almost no research in the subject. After the Great Electric Car Race, I was asked to be the parade master for the celebration in Pasadena. When he talked publicly about the race, [Cal Tech president] Dr. Lee DuBridge always used my name rather than his own in discussing the project. I once asked him why he did this, and he told me, "As the president of Cal Tech, my number one responsibility is fund-raising. That's all I'll say."

What you need to know to understand this is that the oil companies make huge annual contributions. Fred Hartley, Union Oil's president and trustee of Cal Tech, explained that he would tell the school, "We won't be able to continue to make these contributions if you are developing technology that represents a threat to us."

What Union Oil is doing is not illegal. There's no law that says an oil company has to contribute to Cal Tech. Depending on your viewpoint, it's not even immoral.

CS: But it's sickening, wouldn't you say?

WR: Well, the struggle for money and power at the expense of the well being of the common man is certainly an ingrained part of our society. For the hundred years leading up to the development of the internal combustion engine in the automobiles in the early 20th century, our civilization had made great strides in the direction of clean energy. When oil became important, all these efforts stopped.

CS: Where would you personally like to take this at this point in your life?

WR: I'll be 65 soon. I want to visit some labs, conduct some tests—really get involved.

CS: That's terrific. If it's not too personal a question, let me ask you this: When we met last time, you told me that you are a Christian. You said that the Golden Rule commands us to treat others as we would have them treat us—and that this needs to be applied to business, as well as to personal ethics. Yet, you said, you were bothered by hypocrisy. What exactly did you mean?

WR: A great deal of the evangelical world actively works against environmental friendliness. They seem to see fossil fuels as their allies, and ecologically sound solutions as their enemies. I don't see how this is consistent with the idea that all we see around us was the creation of a loving God.

CS: Now I understand. Wally, you're amazing, and I'm proud to know you. Thank you very much.

For more information on this contributor, please visit:
http://2greenenergy.com/renewable-energy-facts-fantasies/.

THE SOCIOLOGY OF DRIVING

One reads a great deal about electric vehicles (EVs) and their potential to provide transportation in a far more environmentally responsible way than the internal combustion engines they will be replacing. In the next few interviews we explore this idea from a number of perspectives: economic, sociological, political, and technological.

By way of introduction, EVs are powered by the conversion of electrical energy stored in batteries and/or capacitors into kinetic energy. As discussed here, there are numerous benefits, but there are also significant challenges associated with the migration away from internal combustion (gasoline and diesel) engines to EVs, which I summarize as follows:

Benefits to Owners—

Lower fuel cost. The average cost of electrical energy is many times lower than the costs of gasoline/diesel. This is due to numerous factors, including the much higher efficiency by which the stored electrical energy is converted to kinetic energy with far fewer losses to heat, etc.

Lower cost of ownership in the long-term. EVs have far fewer moving parts than those based on internal combustion engines, resulting in far less frequent and expensive maintenance.

Performance. EVs have fantastic torque curves, producing far better acceleration than even the highest-end gasoline-powered sports cars.

Challenges to Owners—

Higher initial vehicle costs. Although costs are coming down each month, early adopters of EVs will pay a premium over the equivalent gasoline- or diesel-powered car.

Inconvenience. Refueling EVs means recharging, and, unlike gas- and diesel-powered vehicles, the infrastructure to accomplish this is not ubiquitous. Depending on how you drive and your level of patience, EV ownership has the potential to be inconvenient.

Sacrifices. Though the technology for onboard electrical storage (batteries, capacitors, and fuel cells) is improving constantly, such technology currently forces important trade-offs in cost, energy density (how much energy can be stored in a given unit of battery volume and weight), and cycles (how many times the batteries can be charged and discharged).

Benefits to Society—

Eco-friendliness. There are no noxious emissions from the vehicle itself, and thus the potential exists for extremely clean transportation if the electrical energy is generated from renewable sources.

National Security. As discussed above, it's unclear why a nation so concerned about security and so anxious to stay out of expensive and unpopular wars would not do everything in its power to rid itself of its addiction to foreign oil. And, regardless of exactly where we are with respect to "peak oil," the cost of this addiction is bound to get worse as time goes on—financially, militarily, and culturally—since oil is becoming increasingly scarce.

Challenges to Society—

Ultimately the widespread deployment of EVs will require:

Improvement to the electrical grid.

An increase in the number of charging locations.

Heightened importance of migration to renewable energy generation methods.

Potential shortages. Some people argue that the migration to EVs will cause new types of shortages, e.g., lithium carbonate, the compound from which lithium-ion batteries are made.

I personally do not believe there will be a serious issue here, for what that's worth. We really haven't even begun to look for lithium. We didn't think there was oil underground in any great supply until we started to look for it in the early 20th century. We found it in a big way—unfortunately, we burned most of it up.

I'd like to begin this group of interviews with a talk on sociology, in which I explore my theory that we as a society are in the process of making a quantum shift in our attitudes towards driving. It's well known that in developed countries, most of us view the cars we drive as far more than utilitarian objects. Cars are about the wish to appear affluent, they're about sex, they're about one-upsmanship over our fellows—they're about a great number of things that have nothing to do with the practicalities of moving from A to B. I believe (though it appears I may be alone in this belief) that this is all in the process of changing. I think that a great number of people have begun to realize that every time they put their foot on the gas pedal, they're doing so at the expense of the health and well-being of themselves and everyone else living on Earth—and that this will eventually cause an enormous change in our automotive purchasing behavior.

In order for this to occur, this "we're all in this together" attitude will need to reach a critical mass. But I point out that such seismic shifts in attitude are not unprecedented. Recall what happened regarding mink coats in the 1960s. When I was a little boy, my mother and her friends wore mink when they went out to their parties. A few short years later, anyone stupid enough to have not gotten the message was regarded as a pariah—a Cruella de Vil.

To investigate, I spoke with a great number of people, including Dr. Michael Kearl, Professor of Sociology at Trinity University in San Antonio, Texas.

Craig Shields: Thanks so much for taking my call, Dr. Kearl. It's funny; I thought it was going to be easy to find a sociologist to talk about driving, and the migration to clean vehicles, but it's been much harder than I expected it to be.

Michael Kearl: Glad I could help. For years, I've been interested in the relationships and the social structures for the road—how different communities develop different kind of cultures with regards to driving etiquette, honking and so forth.

CS: Yes, that is interesting. I never knew anyone studied that. But for sure, anybody who's flown into Logan, rented a car, are driven it through Boston is in for a rude shock.

MK: Right. You've got that—and then you've got the whole element of conspicuous consumption and displays even though everyone's in gridlock and you in your Maserati are going no faster than the beat up VW in front.

CS: Let me ask you this: do you perceive a sociological trend in which American drivers wean themselves away from this identification with the car they drive—you know, "I am what I drive?" In particular as it pertains to fuel efficiency? I'm talking about electric vehicles, plug-in hybrids, and so forth.

MK: Here in Texas, to be honest with you, I haven't seen that much. All I see are large new Suburbans and so forth.

CS: OK, but do you suspect there'll come a time when the American psyche will begin to bend on this?

MK: Well probably, if you can get gas up high enough it will have to. But you know I just think that the whole idea of the car and the open road is just so firmly ingrained in our cultural identity in the 20th century that it will be hard to wean. Our car is one of the few zones of solitude we have—next to the bathroom, I might add. And that's why bathrooms have gotten so much larger over time.

CS: Wow, that's really interesting. Well, talk a little bit about that if you would, Dr. Kearl. What do you see as the main sociological trends in driving in the United States at least? In other words, what would you say defines the American psyche vis-a-vis driving?

MK: There's the allure of the open road and the psyche that you talk about varies by age group; I see lots of commonalities between the 16-year-old and the 80-year-old driver. For both, the act of driving is a symbol of independence and freedom. And of course with the aging of society and the need to get the old off the road, that will be one of the dramas that come up in the next few years I think.

There's a sense of entitlement that I have the right to drive and my basic identity when I go try to cash a check, what they want is my driver's license, I mean...how thoroughly ingrained is that?

In terms of that mindset that you are what you drive and the attributions people make of each others' social class and politics and taste and so forth is certainly captured by the various models on the road now a days.

CS: Well, that's my question. You don't see that changing?

MK: No, I don't. The auto's certainly in the top three or four of purchase prices that Americans spend in their life and it's become truly a symbol of self for many individuals.

The entire industry depends upon that. I mean how else can you get someone to spend 50 or 60 thousand dollars and they're on a clogged freeway, doing 20 miles an hour.

CS: Right. As you pointed out, right behind you is a car that's worth $400 and it has dents the size of small refrigerators in it.

MK: You bet—they're going the same speed.

CS: Yes. Well that's very interesting, I was trying to get a hold of the people at Zip Car, the people who do car sharing.

MK: That's a really good idea.

CS: Yes. Some people—and it might be a fraction of a fraction of American drivers—say "You know what? I don't need this anymore. I don't need to make a statement with my BMW. I can make a statement some other way, but I live in Boston or Washington DC where driving is horrible, parking is worse, and to hell with it."

MK: Yes. Well, there's certainly a difference in that East Coast versus West Coast mindset. And given your perspective there in California, there's just so much status involved in that car you drive.

CS: Absolutely.

MK: Think about whole vanity plate—there's another status thing to try to give some identity. And then there's the bumper sticker—that's one of the last free editorial spaces that people have—next to the web. I love seeing someone that's got ten or twenty causes on their back windshield and bumper, you know?

CS: Yes. In other words, if there's any migration it's going to be extremely slow in terms of getting the typical American driver to define himself differently.

MK: Absolutely.

CS: Well great. I definitely will use this, but I have to say that this interview didn't come out the way I thought it would. But Dr. Kearl I can't thank you enough.

MK: Well keep me posted on your progress!

CS: I certainly will.

For more information on this contributor, please visit:
http://2greenenergy.com/renewable-energy-facts-fantasies/.

HYDROGEN FUEL CELLS

I've known Steve Ellis at Honda for a few years now, initially through my partnership with electric vehicle evangelist Bill Moore at EVWorld, where Honda was a long-term sponsor. Steve is one of the most important spokespeople for hydrogen, and I was pleased to get his take on this very interesting and controversial subject.

Craig Shields: Steve you want to tell us a little about the mission here with respect to Honda, fuel-cell cars, and the trajectory of hydrogen, generally?

Steve Ellis: At Honda, my role here today is focused on the hydrogen fuel cell car. But our work with true alternative vehicles started in American Honda in 1996 as we re-formed this alternative fuel vehicle department, to bring our electric car the EV-Plus and our natural gas car into the market. So that was a transitional point when it moves from behind the fence of research and development, labs, technicians, white coats and things like that to the market side.

So, I've been out here in this space for well over ten years, and I've seen this landscape of an alternative fuel vehicle really ebb and flow along the way. When we talk more narrowly about hydrogen fuel cell cars, it'll be in the context of what's happened over the last ten to twelve years.

Back when the California Air Resource Board (CARB) created the Zero Emission Vehicle Mandate for automakers, it was only focused on one thing: tail pipe emissions—not CO_2 emissions, but smog emissions. There was no such thing as a gasoline low emission vehicle back at that time. We introduced it for the first time, you might say, in 1996. There was no such thing as a gasoline ultra low emission vehicle. We introduced the first ones around 99' or so. The natural gas Civic, when it was introduced, was so clean it was classified as just one tenth of that ultra emission vehicle standard, or ULEV, and that was levels that were virtually unmeasurable by California's Air Resources Board. At that time, government's role was really to push automakers to make cleaner cars for one purpose and that was for cleaning up the air, or smog emissions.

That was the issue, health effects of smog. Now if we roll through time here, we had gasoline ultra low emission vehicles introduced, we had gasoline super ultra low emission vehicles introduced to a new standard that did not even exist back when this other mandate was created. So my point is, we automakers were able to accomplish, with gasoline vehicles, something that the Air Resources Board and others thought could only be done with electric vehicles.

That's the preface to our work today at Honda with hydrogen fuel cell vehicles. We ask, "Can we introduce a zero emission vehicle that not only accomplishes that original goal of cutting smog emissions but can cut CO_2—climate change emissions?"

CS: Right. In other words, nobody cared about CO_2 in the late 90's.

SE: Those people that cared were on the leading or bleeding edge of recognizing that there was this connection between man-made CO_2 emissions and climate change. And if you think about it, it wasn't that long ago that the best research scientists came up with those theories.

So that became a new emphasis, and the Energy Policy Act that was created during the (GHW) Bush years, and then carried through into the Clinton/Gore era and now it's mostly not given much attention because climate change has now one-upped that.

The purpose of the Alternative Fuel Act was to focus on bringing alternative fuel vehicles to market, for the sole purpose of reducing dependence on oil. See now here's the second purpose, you had the smog, now your dependence on oil. Well along came, Al Gore with his movie, *An Inconvenient Truth,* and woke up America about this issue of climate change.

Hydrogen fuel cell vehicles can accomplish 100% of that original goal of zero tailpipe emissions. It can accomplish 100% of the issue of dependence on oil, because hydrogen will always be a domestic resource. So then, the third part is the CO_2. And that's where most of the emphasis is today.

CS: I wrote a blog post about last fall's "The Business of Plugging In" show where General Wesley Clark said, "Energy represents a national security problem." That's news? The DOE was formed to deal with this problem in 1977. Hasn't it been clear for 30 years?

SE: I think it's been clear to a few. And I would say that Honda is one of those few that understands, that without a sustainable business model, sustainable on the energy side, there is no business. Over the years, it's been said that Honda seems to have more long-term vision than most in the automotive arena.

The people I know at the energy companies that are truly dedicated to this cause of reducing dependence on oil, or helping to resolve CO_2 issues; they are very sincere. It's from the heart; I don't get the sense that it's just window dressing.

CS: Ok. I'm sure there are some good people in these companies.

SE: Tragically now we've had this huge economic impact, and that changes all the equations. If a company is very profitable year after year, it can invest in research and technologies with a longer-term vision. If not, you have to put every resource into what is going to give you a return in the short-term.

That's the challenge of economic conditions.

CS: OK. I'm personally not convinced on this, but no matter.

My primary reason for being here this evening is to talk about hydrogen. I think electric transportation enthusiasts who, most of whom, have seen *Who Killed the Electric Car?* believe that hydrogen is a red herring. Physicists are generally looking to battery technology as opposed to hydrogen because of the efficiency issues. By the time you electrolyze water, store the hydrogen, put it through a fuel cell...you're at a fraction of the efficiency of a battery and an electric motor. And I'm sure you're hit hardest on the infrastructure issue. There are plenty of places to charge your car—and building more is fairly inexpensive. This is very clearly not the case with hydrogen.

SE: Sure. So, we've covered a little bit of background. You had asked me about the most basic aspects of a hydrogen fuel cell vehicle. These are basically taking hydrogen, storing it in a tank of a vehicle and when you introduce it to a fuel cell, which is called a stack, one side of this plate you introduce hydrogen and on the other side you introduce oxygen. It's called a PEM (proton exchange membrane). And when that exchange occurs it creates electricity. So you think of it as each plate maybe as like one volt, so if you want a 220-volt fuel cell stack you have 220 stacks. So in that regard it's not unlike a battery.

CS: Exactly. A lot of people go "oh, I want an electric car" as if a fuel cell isn't electric. I'm sure you've noticed that people don't understand this.

SE: Yes, that's a good point. So a fuel cell vehicle is an electric car, period. That I think has to be clear. It is everything that a battery electric car is except for the battery. In lieu of the battery it uses this fuel cell stack to provide the electricity to the electric motor. It makes that electricity by storing hydrogen on board and introducing hydrogen and oxygen from the air, in that stack, and that's what makes the electricity. The only by-product is water.

CS: Yes, exactly. It's because that one electron in the hydrogen atom wants so much to join that oxygen atom to make water.

SE: Exactly. You mentioned the movie *Who Killed the Electric Car?* and you know, I met the producer, I know him, we have good dialogue, but very frankly, the movie was made for one purpose: to sell a movie. It's like various documentaries—it's crafted in a way to cause interest and cause people to come see it or sustain DVD sales or whatever—hence, to make money. What's missing in the movie is a lot of pieces that if combined together would be a different truth.

CS: More fair perhaps?

SE: That's right, more fairness. The point you made about the movie insinuating that this is just a red herring, is simply not true. It's just plain false. If there's people that don't believe an automaker has the best of intent in their activity, I'm not sure I can help that. All I can do is say, from Honda's standpoint, we have a track-record of performance based on action in engineering work by a team of very brilliant engineers. Honda's reputation includes not pursuing things that we think aren't sustainable or in the best interest of society.

I'll give you a simple example, Craig: the ethanol vehicle in America. We watched that carefully. We know technically how to do it; we understand what technology is needed to be added to the car to make a car run on this mix of anywhere between, 0% ethanol up to 85%. But our company made a strategic decision to not do it at a time when many others did for purposes that, let's say may have had to do with meeting CAFE standards, or to be perceived as playing in the arena of offering alternative fuel vehicles that were supposed to have a certain effect on dependence on oil. But, we felt all along that if as long as it was a food-based ethanol, it probably would have some other societal harm, some unintended consequence. We just chose not to do it.

CS: Good for you.

SE: Now take what I just said and align that with what we're doing with hydrogen fuel cell vehicles. We see, not only that it has sustainability, but it also provides a significant value to mankind. And in any time we have pursued that Craig, think about it, it's the pursuit of not just a single goal. Think about those three points I made; smog emissions, dependence on oil, and CO_2. Think of it as three legs of a stool. And we want the stool to be solid and to stand on its own. So, those are the technologies that we're going to pursue.

I don't want to come across that Honda has some beef against E85. What we have is the support for technologies that we think are sustainable.

CS: OK, Steve. I'm sure readers will presume you have your heart in the right place; I'm happy to make that presumption, at least. That's doesn't mean it's a good idea in terms of business and technology.

SE: Right. That was going to be my next point. So, when you read the blogs, when you listen to what people say, when their interest is to negate or harm hydrogen and fuel cell vehicles. 99% of the time it occurs in the context of versus battery electric vehicles. I think that's where the first flaw occurs. They should be compared to internal combustion engines.

CS: Oh, I don't think it's unfair for them to do that. Progressive people want to make a change. And the question is: what do you change to?

SE: Let's be clear. You know, electrification of vehicles starts at the most basic, let's say just a start/stop version—all the way up to a battery electric or a fuel cell vehicle. Some would have you think that we can jump right from where we are today—98% dependence on internal combustion vehicles, even if you bring hydrogen to the mix—all the way to that pure electric power train.

Keep in mind that this is the 10th anniversary of the first hybrid electric vehicle being sold in America, and they're just starting to approach 2% of all vehicle sales. See what I'm getting at?

CS: So your point is that the EV adoption curve is slow, thus we need hydrogen? I don't see this.

SE: You don't have to look at one in lieu of the other; that's where the huge flaw is. You have people working against hydrogen, and you have (U.S. Energy Secretary) Stephen Chu making this announcement. You don't have the National Hydrogen Association, the US Fuel Cell Council, California Fuel Cell Partnership or anyone involved with hydrogen fuel cells saying "Don't fund batteries." And that's the flaw, Craig. There needs to be both; they're complimentary of one another, F6 Clarity advances every technology a battery electric vehicle person would want to see advanced—plus hydrogen fuel cells.

So why would you shunt that to the side of the road just because your myopic view is solely fixated on batteries and charging? Today, we're already proving that the hydrogen fuel cell vehicle has this capability, driving from the border of Mexico all the way up to Vancouver, over the course of about five days, fueling along the way. If you lined up battery electric vehicles at that same starting point, how many days later would they have arrived?

Now let's get to the science of it. Let's get to that point you made about efficiency. The goal is lowering dependence on oil, zero smog emissions, and CO_2. So with the hydrogen fuel cell vehicle, what do you get? You get 100% of that goal towards your vehicle emissions; you get 100% toward that reduced dependence on oil, and just like a battery electric vehicle, you get a significant reduction in CO_2 emissions.

Now here's why I'm putting it that way: electricity done wrong, with an older, inefficient coal fire power plant, ends up having that vehicle emit more CO_2 than a gasoline vehicle.

CS: I've always read that this isn't true.

SE: It is. I'm glad you're saying that frankly because I hear others out there saying it too. I'm not calling anyone a liar, but charging battery electric vehicles with electricity done wrong nets you worse CO_2 emissions than a traditional gasoline vehicle. The well to wheel studies under the GREET model proves this. That's the unequivocal method developed by Dan Santini at Argonne Labs; it's all public information. That's the model that everyone uses in government industry science to evaluate this well-to-wheels value on CO_2 emissions for vehicles. UC Davis, UC Berkley, UCLA, and UC Irvine, all of their researchers that are dealing in this arena understand that clearly. And California has put it in the light better than any other state because, guess what, this is the state that has now passed legislation for LCFS, Low Carbon Fuel Standard.

Each state, of course, is different re: how electricity is generated. I'm not going to sit here and say "It doesn't make sense to plug in battery powered cars," because in states that have a cleaner energy mix you get a net CO_2 reduction.

So, let's connect that back to hydrogen fuel cells. A battery electric car has zero carbon emissions if you charge it solely with zero carbon electricity—solar, wind, hydroelectric, geothermal. Just like a hydrogen fuel cell car.

CS: I see your point, though I'm not sure how powerful an argument it is. If you want X kilowatt-hours of energy in your tank, you're going to start with four times as much energy for a fuel cell car as with a battery electric. I suppose if you contemplate a world with an infinite supply of extremely inexpensive renewable energy...

SE: Let me take you through this briefly. You'll read it on the blogs and hear people say "Yeah, well what good is hydrogen if you make it from natural gas, because it still makes CO_2 emissions?" What they fail to say is: "Just like electricity makes CO_2 emissions unless it's 100% renewable."

So let's talk numbers now. Take the F6 Clarity with its tremendous efficiency—over 60% efficiency tank to wheels. It cuts CO_2 emissions 60% when you make hydrogen from natural gas—and it's getting better. A natural gas vehicle, like the Civic GS cuts CO_2 25%.

Compare that to the inefficiency of internal combustion, which is 18% - 20%. So, when you combine an extremely efficient vehicle with the efficiency of reforming natural gas into hydrogen and running it through a very efficient vehicle, you cut CO_2 60%.

CS: Versus internal combustion?

SE: Yes exactly, I'm sorry. Versus internal combustion comparable size and weight class vehicles.

That's the key Craig. Now, how do you get to 100%? Renewable hydrogen. So if you make the hydrogen from bio-methane—if you're capturing for the bio-methane and using it in a very efficient manner. You're creating it from something that otherwise was waste. So how do you get 100%? Electrolyze water from solar.

CS: Right. Or wind or whatever.

SE: Right. You see what I'm getting at? A hydrogen fuel cell vehicle can be a zero CO_2 emission vehicle—just like a battery electric vehicle. A battery electric vehicle advocate should be saying, "Battery electric vehicles can be zero carbon emission vehicles...just like a hydrogen fuel cell vehicle can be too." You see the point? That's what's missing in this debate.

CS: Thanks so much, Steve.

For more information on this contributor, please visit:
http://2greenenergy.com/renewable-energy-facts-fantasies/.

TOUGH REALITIES

ADVOCACY, ECONOMICS AND OTHER ISSUES

ADVOCATING FOR ELECTRIC VEHICLES— PLUG-IN AMERICA

When we think about advocacy for electric transportation, Plug-In America is the first name that comes to mind for most of us. Spokesperson Jay Friedland talked to me about the organization's activities and the numerous thorny challenges it faces in the real world of politics.

Craig Shields: I've seen *Who Killed the Electric Car?* and I've spoken with people who were in the ZEV (zero emission vehicle) Mandate meetings in California. I see some level of evidence that there is an "unholy alliance" between the oil companies, who obviously have a motive to delay the advent of electric transportation, and government regulators, who are supposed to be acting on behalf of everybody.

Jay Friedland: Right, it's both the oil companies and the automobile companies—although the automobile companies are changing, having gone through the sort of near-death experiences—or in some cases, death experiences. But fundamentally what happened was that each of these entities was really trying to preserve the status quo. The oil companies, for instance, are now the largest supporters of the hydrogen economy. They see hydrogen as a mechanism for them to continue to have a service station—to continue to provide a consumer with something they can pump.

Electricity, on the other hand is ubiquitous and the world is moving to a paradigm where you're much more likely to charge your car at home overnight. The way I put it is that you're going to charge where your car sleeps. When your car is not in use—at work or at home, that's when it will be charging. Then it'll probably be at your convenience—at a supermarket or a Costco or a theater or something like that.

You asked about regulators—there is no doubt that they have gotten too cozy with the people that they regulate. Probably the best example of this was in March of 2008 and the Zero Emission Vehicle Mandate; it was really the perfect time for CARB (the California Air Resources Board) to act. Oil prices were rising, market forces and consumer demand for electric vehicles was building; there were so many things converging. Auto companies were signing up to build either plug-in hybrids or true electric vehicles. And at that moment in time, CARB decided to continue to eviscerate the ZEV mandate, cutting it by 70%, so that certain automakers could continue to do fuel cells. Most notably Honda and Toyota were the primary people lobbying for this.

At a time when they could have held the industry's feet to the fire, they abdicated their responsibility and cut way back. And they tried to make it look good because the staff had recommended a 90% cut in the requirements and they came up with only a 70% cut, but the bottom line is there were 70% lower numbers than they had themselves decided at the previous round of those talks in 2003, where they had cut back dramatically. The regulators have shown no force of will about this.

But interestingly, in the meanwhile, the market opportunity has changed so dramatically that we now have car companies like GM and Nissan and Ford saying they're going to produce hundreds of thousands of these vehicles—without regulation involved. I mean there are regulatory incentives with the new CAFE standards, but the bottom line is that they're looking at producing far more vehicles than are required by the regulations, because the market wants it.

From a policy standpoint, Plug-in America is moving further away from a CARB-mandated mode, and much more toward creating the incentives for both manufacturers and consumers—that's why we pushed and educated and drove so strongly for inclusion of what ends up being about $14 billion worth of potential incentives. We have to build the factories and get cars out there and get everything in place, but in terms of tax credits alone, now electric vehicles are eligible for up to $7500. We got a law changed regarding plug-in hybrids, since they have batteries in them, and now instead of being a cap of 250,000 for the entire industry, it's 200,000 per manufacturer. So each manufacturer coming in has the ability to offer incentives for 200,000 vehicles.

There are details that are confusing, but at a high level, we now have ten auto manufacturers that have all committed, and that means 2 million cars that are eligible for that $7500 tax credit. So this is a really significant public policy. We've worked very hard with CARB—actually in a positive way—to get them to do additional rebates. They haven't put enough money into the program, but at least we've got the marker on the books and so the first several hundred Nissan Leafs will get a $5,000 rebate on top of the $7,500. So if you believe Nissan will come in at $30,000, and you get $12,500 in incentives, that brings it down to $17,500—that's an affordable electric car. That's deal-changing in terms of the ability for a consumer to look at this and say, "Does this meet my needs?"

CS: Let's look at this in terms of the history. The way I remember it, CARB tried to force EVs on the car companies in the late 90's, but the OEMs said, "Make the mandate go away and we'll do this on our own"—which of course they lied about. But now it looks like this is happening through market forces as opposed to legislation regulation. Isn't that good?

JF: It is good. I think that what you need is a balance; it's like so many things. The auto companies would not have put airbags in the cars; they would not have put seat belts in the cars; they would not have put anything that we think of today as being incredibly important safety factors. Without the CAFE standards notching up and the auto companies having to think about how they are going to meet this, it would never happen. A week after the CAFE standards got announced, everyone was saying there was no way we're going to be able to meet these aggressive new numbers, Ford announced this ecoboost technology—an enhanced turbocharger which will give them about 25% better gas mileage. They announced that a week afterward, meaning that they had that technology sitting on the shelf, ready, in case this thing needed to happen. The same thing with Toyota, again to protect their market position around hybrids, has been bad-mouthing plug-in hybrids and electric vehicles—very publicly, in a lot of different forums. And that's because they own 75% of the hybrid market. They, in essence, helped create the regulations that caused hybrids to come to dominance in fuel-efficient vehicles.

As I look out to the future, I think we'll have a combination of technologies that deliver these high-mileage vehicles—with the highest ones having the highest component of electricity—that's the clear path. But it will take us a little while to get there and we need to do everything we can. So it's a balance between regulations, incentives, tax credits, and things like that—overall, different public policies.

For example, look at major urban centers, congestion zones. Dealing with a congestion zone is a phenomenal way to improve urban air quality—you need the kind of public policy, like London has, that is very restrictive. You have to pay large amounts of money to want to drive into the city of London unless you're a zero emission vehicle; then you can drive in at no charge. Tomorrow is the first of the Senate hearings on the next round of electric vehicles and plug-in hybrids being able to use the carpool lane.

CS: I want to go back to what you believe to be the real reason that EVs are slow to come to the market, if I could. I was at the "Business of Plugging In" conference in Detroit last fall and Toyota stood up and talked about how they were bringing along plug-in hybrid technology and the MC said, "This is brave. This is remarkable." I almost threw up. They could have done this years and years ago if they had wanted to or needed to. If they weren't already perceived as being "green" and didn't have an existing and profitable hybrid platform to sell and protect, they could have done this thing a decade ago.

JF: Absolutely. I have a Rav 4 EV; I've had it for eight and a half years—and indeed they could have ramped that car and made second and third generations. The GM EV-1—same thing—they could have ramped that car. But you have to remember the environment of 2001, 2002, 2003—cheap gasoline, really the only public policy that we had around climate change was fighting wars to protect oil supplies.

Honda was really the leader in saying, "No, we don't need to do battery electric vehicles or plug-in hybrids. Existing hybrid technology will take us to 2015 or so when we'll be able to commercialize fuel cells." What that neglects is the hundreds of billions of dollars of infrastructure that would be required to deliver hydrogen, and the fact that hydrogen is about a quarter the efficiency. If I'm generating hydrogen and I wanted to do it in a renewable way, it would take me four times as many solar panels to make the electricity as it would to just put up one set of solar panels to make the electricity to just put into a battery.

This is where Toyota and Honda are going to be potentially against Nissan and GM; we're seeing a schism between the automakers, which is very interesting and important to the long-term. For example, Nissan just announced at their LEAF announcement on Friday that they have numbers that say it takes about 7.5 kilowatt-hours of electricity to refine a gallon of gasoline. And that same 7.5 kilowatt-hours will drive a Nissan LEAF 30 miles. So, if you had a 30-mile-per-gallon vehicle, not only does it take that gallon of gasoline, but it also takes the 7.5 kilowatt-hours of electricity. So, in essence, you could stop refining that gasoline and have the grid capacity to feed the battery cars.

I'm incredibly impressed how committed Nissan is—how far they've gone in going up against their existing business, which is gasoline cars. Fundamentally, they're confident. They know that they're going to keep selling gasoline cars for some period of time, but they believe that we've got to change this.

The argument was: The fuel cell vehicles' fuel pods are $500,000 to $1,000,000 apiece, so we can't ramp the volume you want us to. Nissan came in and said. "We'll do it with battery electrics." Toyota and Honda said, "No, no, battery electrics are a bad idea. They're always going to be prohibitively expensive, and we can use fuel cells and drive the price down—from half a million dollars a car.

Keep in mind that the cost of batteries is falling by a half about every three years. So we are rapidly traveling down a curve where the battery costs are getting lower and lower.

CS: But let me ask you about charging infrastructure. The people at Honda—and I know them personally—argue that hydrogen has a place in transportation precisely because of the fact that neither fast charging nor battery swapping are acceptable options.

JF: But million dollar hydrogen stations are? Sorry, but the Honda people are being totally disingenuous.

Electric charging stations are not terribly expensive. Fast charging could cost about $50,000 a station, as most gas stations are pre-wired with 480 and have three-phase high power for a variety of different reasons—it's already there. Hydrogen stations, and there is no economy of scale with this, are a million dollars apiece. So you can put twenty electric fast-charging stations in for the price of one hydrogen station. The price tag is about half a billion dollars.

CS: $50,000 into half a billion dollars is 10,000. Is that enough?

JF: Yes, well, every 30 miles on the interstate highway system.

Here's the other thing: 90% of the travel is for commuting 40 miles or less. I was at a green show in San Francisco and someone walked up to me and, said, "Well, I don't know if I could ever have an electric vehicle. What if I want to drive to Tahoe?" It's about 200 miles one way, and so I asked, "Well, OK, how often a year do you go to Tahoe?" And the guy says, "Well I've never been to Tahoe, but what if I wanted to go?" We call those aspirational trips.

One of the sad things is Chrysler announced that they were pulling out of the EV industry after the deal with Fiat, and Chrysler was working on a Voyager Minivan EV—a perfect vehicle to drive around town.

My guess is that pure EVs will represent between 10% and 20% of the market in the future. They're probably not for everything unless we see larger changes, but what about plug-in hybrids? For our family, the perfect set of vehicles would be a very high mileage plug-in hybrid and an EV.

I also know a lot of people that have a single car—an EV—and it works just fine for them. They live in urban environments; they rent a car the four times a year that they need to drive from San Francisco to LA, and it turns out that if you think about wear and tear, maintenance, insurance, all of those other costs of keeping a car, it's not worth it. Another example is that I don't own a pickup truck. Probably like three times a year, I either borrow one from a friend or I go down to Budget Rent-a-Car, rent one, and I use it to go to the dump.

CS: This whole subject of the paradigm shift of the way we Americans drive is interesting. My theory is that we are increasingly reluctant to define ourselves in terms of what we drive. We're less likely to say, "Oh, I have to have a BMW because I'm an executive" or whatever.

JF: A lot of people's thinking and behavior is driven by misinformation. You know there's been this whole issue recently where the level of people that believed in climate change has down by like 10 or 15 percent at this point. I find that a little bit weird. How could you believe something and then stop believing it? What does that mean?

I have a friend who teaches an environmental sciences course at a local junior college, and her husband came in to give the talk on climate change—he was paid by the oil companies. He's a denier—"There is no global warming." The university didn't even know that they were bringing him in, and paying him to do this. One of my friends who's an astrophysicist took him apart. So what happens at the end? The astrophysicist got yelled at for not being "collegial" by the dean of the university! What are you, crazy? The guy's spouting lies and I can't correct him? "Oh, well we were trying to be balanced."

CS: Amazing. What are some of the other Plug-In America platforms?

JF: Well, one is that being green can save you money. Being green is about being more efficient. That's why I love this concept about oil—you know gasoline is in essence dirty twice, it's double dirty, because there's an environmental consequence to that electricity that goes into refining it and then you go put it in your car and you burn it and you put twenty pounds of CO_2 in the atmosphere for every gallon you burn.

Even if the Obama administration hasn't accomplished as much as we've wanted them to, from a plug-in car perspective they've done a phenomenal amount of getting the job done. Most of it was because they got the key provisions into the stimulus, which is the most major thing that they've passed. And it'll be very interesting to see how much compromise ends up happening—and that's our next great barrier that we're working on. There are some great plug-in provisions, but if the overall bill is crap, we run into that problem.

CS: My personal theory is that until and unless there is meaningful campaign finance reform, nothing is going to happen here. The only reason these people get and stay in office in the first place is simply because they're taking money to forward things that create profit for a very few people.

JF: I tend to agree with you—and the example that I keep pointing to is hydrogen—it's a gigantic obfuscation. Hydrogen is incredibly well financed; there's a Hydrogen Caucus in Congress—a group of senators and members of the House of Representatives, that in essence meet to talk about how they can advance hydrogen—and that effort is funded directly by the oil companies.

CS: What else are you folks at Plug-in America working on right now?

JF: Imagine you go out and buy a car today. You walk into a car dealer, you haggle or maybe you don't, depending if you have pre-cut the deal. Maybe you've gone online and you've done the research, made your decision, you go in, and that's when you pick up the car. If the car dealer's nice they filled up the tank and you drive it home and a few days later you go to a gas station, you fill it up, and this has been the conventional mode. Well, now imagine it's an EV. For the first week or two you charge on a 110-volt outlet. And then you realize, well gosh, you'd like to charge this a little faster. And then you realize you have to install a charger so you have to call an electrician, pull a permit, get the charger, call the utility and get a better rate for your charger. We're working with the auto companies and the utilities and the municipalities so that we streamline these processes. So you'd go to Nissan, you figure out the car you want, and then you go through this little checklist so the day you show up at the dealer, you've already got your charger installed. So you go home and then you never go to a gas station for the life of that car. But you have to do up-front work; that's not the typical car buying or American consumer habit. I liken it a little bit to buying a home theater system today. It's like you go to Best Buy and you've got the Geek Squad that comes help you install it. Incentives are good, but we have to get rid of the bricks in the road that may slow down the adoption process.

CS: Great stuff. Keep up the good work. I really do appreciate your help.

JF: Sure, my pleasure.

For more information on this contributor, please visit: http://2greenenergy.com/renewable-energy-facts-fantasies/.

ELECTRIC TRANSPORTATION AND ITS IMPACTS ON OUR POWER GRID

The Electric Power Research Institute (EPRI) is an independent, non-profit company performing research, development and demonstration in the electricity sector for the benefit of the public. EPRI's broad array of collaborative programs focuses on the many specific technology challenges of helping its members provide society with reliable, affordable, and environmentally responsible electricity. Mark Duvall is an extremely well respected specialist in electric transportation and the impacts that it will make on our grid. I've met him at numerous conferences and he was kind enough to speak with me for the book.

Craig Shields: Mark, thanks for your help here. I know you've taken a controversial stand on some issues regarding electric transportation.

Mark Duvall: Yes. Right now people say things to get in the press and someone will go out and say, "The San Francisco Bay area needs $1 billion in public infrastructure." And I think I've called that "crazy" a couple of times. People are pitching this as a chicken and egg problem, and I don't think that it is.

CS: I completely agree with you, for what it's worth.

MD: I think if you look at where infrastructure is going to be the blocking point, it's going to be at home. Early adopters generally live in typical residential situations with parking spaces. So the typical person is going to say to himself, "What do I really have to do to incorporate this vehicle into my life?" Early adopters are going to take the plunge no matter what, and that's either going to be a positive experience or a negative experience at the home. They're going to tell everybody because that is what influential people do, right? You wouldn't be influential if you didn't tell everybody about your new gadget. So we really need to work very hard to make that initial experience a good one. That really has nothing to do with the city of San Francisco putting in public chargers.

That said, I basically told every utility that would listen that they need to consider either putting in or working with one of their cities to put in public charging infrastructure. They need to ask themselves that question: "What should we do?" Then answer it. If the answer is, "We can't do it because our regulators won't let us," or if the answer is to put one out in front of their headquarters, that's fine too. But the point is to do something, be out there, learn from it, and then decide what else to do. Show some leadership.

So it's not that I think that we shouldn't do public infrastructure; I just think that we need to look at it on the scale of what is really necessary. We also need to take into account that there are some very large public infrastructure demonstrations being funded by stimulus money. And I think we need to see how those work out—especially with regard to some of the more exotic things like fast charging.

So, I am a champion of residential infrastructure. I think I may be the only one, other than the auto companies, who are also aware that this is what has to happen. I believe, there is going to be a natural balance between plug-in hybrids—or range-extended EVs, as Chevrolet would prefer we call the Volt—and (pure) electric vehicles.

I think it's fundamentally not a great idea to try and fit the attributes of a pure electric vehicle to a very much wider set of customer requirements through external infrastructure. So, in other words, build electric vehicles and sell them to the people that can really use them. Support them with an adequate infrastructure, but don't force the issue. Don't try and force one set of vehicle attributes on all drivers.

Compact cars in the Prius class is where everyone's fighting it out—and you can do a lot with that category; it's a useful vehicle. Yet, while a 24 kWh battery would probably make a reasonable EV, you could make a reasonable plug-in hybrid with as little as like a six or seven kWh battery, or you can go to something with a lot more electric capability, like a Volt.

But the point is that we are going to need trucks and vans and big cars and we are going to need a lot more vehicles. And that puts a lot of upward pressure on the technology to handle these other applications. The nice thing about a car is that it weighs 3000 pounds with no one in it and you fill it full of passengers and it weighs, say, 3700 pounds. That's a little different from a pickup truck that might weigh 5000 pounds and then you hook up an 11,000-pound trailer.

So as you go up and you try and pick up the larger vehicles, which we still buy and drive in very significant quantities, you need to be more clever. You need more plug-in hybrids, you need more adoption of new systems like parallel hybrids, or like GM's dual-mode hybrid which was developed expressly to allow for big trucks and SUVs to do all the things that they do with little to no compromise. So those vehicles can all be made plug-in hybrids.

EPRI has vehicle programs that go all the way up to almost 20,000 pounds gross vehicle weight. So you can plug all those vehicles in, you just have to be very strategic about how you do it, and not try and fit all things to all people. I'm very skeptical of somebody trying to make, say, a battery-powered full-sized pickup truck.

CS: I agree. So you see, then, a reasonably long-term trajectory for plug-in hybrids. In other words, you don't see this as here and gone in a few years?

MD: No, and the simple matter is that, just remember, plug-in hybrids are more efficient in electric mode than electric vehicles are, since the battery is not as big. At the end of the day, you can create a plug-in hybrid that satisfies a lot of people's needs in a midsize car with a 6 - 8 kWh battery. You need a minimum of 24 in a pure electric vehicle. GM is going conservative in the Volt.

You'll always have a smaller battery, therefore they are always a little more efficient because they weigh less. And we've gotten that sort of turned around recently. I have 15 years of experience working with these plug-in vehicles, and the last couple years was the first time I ever heard anyone say that electric vehicles would be cheaper and more efficient than plug-in hybrids. That's generally not the case. So, to be purely accurate and less controversial, what I would say is that in the use of electricity, plug-in hybrids are generally at least as efficient as electric vehicles.

CS: Well, isn't that driven by the fact that right now, the energy density is X and the cost of lithium ion is $Y per kWh? In some conceivable future time, the calculus could have completely changed, isn't that true?

MD: In order for the calculus to change, more than a couple things have to happen. Electric powertrains have to be cheaper per kilowatt than gasoline combustion engines. And they're not. So they have to be cheaper for equivalent performance. That's not strictly per kilowatt, but for equivalent performance.

In other words, you have to get dramatically cheaper to overcome those cost differences. I think it's safe to say that the theoretical minimum for a lithium ion battery is going to be about $200 per kilowatt-hour. Even futurists should not really expect them to get below that. And that is not in pack form; that's not the added cost. That's not with all the vehicle markups and all that. What you're going to see for the foreseeable future is vehicles are going to come out and they're going to cost a certain amount and if the costs of things like batteries and electric drive-train components decrease, automakers are going to try to make money. And so that's going to blunt the impact of decreasing battery cost.

I hope durability increases to the point where people become more confident that the battery they buy today is going to last the life of the vehicle, but at the end of the day I think it's a long way off where an electric vehicle with a full-sized battery is cheaper than a hybrid or a plug-in hybrid.

CS: Okay, I understand.

MD: In the near-term, you're going to see all kinds of things. Because they're not going to be based on companies' making profit. They're going to be based on near-term business cases designed to get their product out in a smooth fashion with no hiccups or recalls or any technical glitches, and to minimize their per-vehicle loss. That's the near-term. In the long term, I would find it very difficult for pure electric vehicles, even in the small sedan, to be cheaper than plug-in hybrids. And that will be especially true as you get to larger more demanding applications.

CS: Okay. I want to come back to this issue of the business case for the EV OEMs, but as long as we are on the subject of infrastructure, let me just round off that part of this discussion by asking about this. Obviously, if you contemplate ubiquitous fast charging or a battery swapping—a Better Place type of idea, then you've got a different calculus as well. That changes the amount of onboard energy storage that is required for most cars. I say "most" cars, because if I have a LEAF or whatever, I'm not going to take it on long trips, because we're a multi-car family. But if I want to take a car from here to go skiing or something like that, it does require either onboard gasoline or the contemplation of something like fast charging.

MD: If you want to drive a pure electric vehicle between cities, you either need a battery exchange or some method of recharging the battery in about the same amount of time it takes to stop for gas.

CS: Precisely.

MD: That is the minimum that consumers will accept. If you think that's a good idea, when's your next long trip coming up?

CS: Actually, in about two weeks. I'm going from here to Utah. It's a nine-hour drive.

MD: What kind of car do you have?

CS: An aging BMW 540.

MD: OK, well, so to test out my thesis about the driver experience, I want you to start with an empty tank. I want you to go to the gas station when you start your trip and I want you to limit yourself to no more than two or three gallons of gasoline. And I want you to stop every time you need gas. That's probably overstating it, because with 100-mile range you would start to have range anxiety at about the halfway mark—at a gallon and a half. But limit yourself to 2 to 3 gallons of gas for the whole trip. And then we'll have this call again.

CS: Well that's my point. The last damned thing I want to do. I'd rather ride my bicycle to Utah than do that.

MD: First of all, I am not actually certain that battery swapping is a feasible business case. I know a lot more about fast charging. So the business case to me for all publicly used infrastructure is uncertain to me. I don't think I've really seen numbers that I can really say, "Yeah, there are people that can run businesses doing this and make money at it." I've seen the Coulomb (EV chargers) business case and I think that's a very telling indication from people who have thought a great deal about this business of how you would manage if you wanted to operate charging networks.

You just can't be on the hook for the installation, maintenance, or capital costs. So, they sell the charger and they operate the system on behalf of their customers. Who are the people that buy, install, and maintain the equipment? So, if you had to buy, install, and maintain the equipment, and operate the network yourself, I'm not sure what the business case is for you; I'm not sure how you make money doing that without charging very high prices.

I'm writing a big paper about this for one of our utility members, so this is a big discussion. If you want to drive between cities, you need some way of instantly swapping the battery. However, people already hate going to the gas station. So it's kind of a stopgap measure at best. Or you put in really big batteries and you do some combination. Like driving a Tesla on your trip wouldn't be as bad because it has roughly 200 miles of range. A guy like you with a BMW 540 is probably not going to get 200 miles, you'll get like 150. So you'll stop every couple of hours and recharge. So that's probably not as bad. That would be really acceptable in Europe where, I have in-laws from France and they practically pack a picnic lunch for a two-hour drive and they're going to stop halfway. Or you're taking trains, or you're doing other things.

But let's stick with the US. You either have five-minute fast charging or you've got to have a plug-in hybrid. Or you've got to have a conventional car that you rent. You've got to use something else…. an electric superhighway concept. People aren't really stopping to use the equipment that much. You would have a hard time making money doing it. You'd have a hard time getting power to some of these installations. Those gas stations that are out on the off-ramp, they are feeding off of a long circuit; it's miles from a substation.

So, the utility may be looking at a whole line upgrade just to get you the 500 kW you'd need to run a station like that. But let's assume that's not an issue. The point is that you're driving, you're pulling over every hour to recharge your car.

Or you move these technologies into the city and you use them in two ways. For people who want to drive an electric vehicle and it does 90% of their daily driving but they are afraid to get caught without a charger, they're afraid their kids will get sick and they'll have to run home before they'll get a chance to fully recharge, or some unforeseeable but rare circumstance. So then, the fast charging network or the battery-swapping network has a role to play. It also could have a role to play for people who don't drive a lot, but who don't have a parking space. So if you don't have a parking space you could go and get your car refilled at these battery-swapping places or a fast charging place where you just pull in for five minutes.

But I want to point out that the 2 to 3 gallon rule applies. People that have to do that every day aren't likely to be happy with it. People who live in the city and don't drive their car to work or don't drive very many miles could be relatively happy with that idea. Or they could just buy a plug-in hybrid. Either way, the point is that the cost of that infrastructure would greatly exceed its benefits if it were not done correctly. I haven't seen the numbers, but my preliminary back-of-the-envelope calculations show that these are very tough business cases—especially with a network, because by definition a network has to account for the fringe.

Regarding fast-charging, Tokyo has a nice demo. They had a charger in the basement of their parking garage in their Tokyo headquarters. And they had some electric vehicles for their employees to drive that they could all fast charge. And when I mean fast charge, 50 kW. This is not interstate fast charge; this is five minutes that will get you the rest of the way home or a half-hour will get you pretty close to a full recharge. So this is not, I would say, super fast charge.

And before that, when they just had one fast charger, people drove very limited routes and always came back with more than half the battery capacity remaining. Because they were worried about getting caught. But when they put a fast charger on the other end of town, people drove everywhere, and tended to come back with much less than half the battery remaining. So the point is that putting that charger out there created a simple network, and caused people to drive a lot more and with a lot more confidence. But they never used that charger. So if you're the guy that owns that charger and paid $30-$40,000 for it, and $20,000 to put it in the ground and wire up electricity to it, and has to pay the peak demand charges to operate it sometimes, you're sitting pretty unhappy.

In the meantime, cities—communities in general—should focus on creating practical public charging networks that serve the vehicles that are in their community today—meaning, you get ready for a rollout and you have a comprehensive plan, but you build to suit. So, for the Bay Area of California, step one is refurbishing the existing infrastructure. Put new charge stations in on top of the existing wires so you have the modern connector which probably won't even be available for another two or three months, and get ready. There are thousands of stations out there.

The other thing is that, I think that there are decisions to be made as to what people buy. So if you have a choice between a vehicle like the Chevy Volt or a pure electric vehicle, you really should take a look at your lifestyle and not depend on a public infrastructure to get you through the day.

One of my closest friends is a guy named Dan Santini who is an economist for Argonne National Labs and he's been doing these studies with EPRI from the beginning. He would look at this and he would say we need to do plug-in hybrids with the smallest possible battery because that gives you the greatest benefit per kilowatt hour of battery or per extra vehicle cost.

Obviously it's a combination of economics and driver demand and what makes people feel good about themselves. Chevrolet certainly didn't intend for the Volt to be the most cost-effective plug-in hybrid. They intended for it to be a vehicle that would meet most of their customers' needs, leverage powerful technology that they had already developed in-house. It's basically an electric vehicle for the first 40 miles of the day, yet it can do everything you need. Nissan has got their approach. Who knows, five years from now, Nissan could also be making a plug-in hybrid and GM could be making an electric vehicle. We don't really know where this is going to go.

There are enough early adopters to buy all these things. I don't think demand is going to be an issue. Supply is going to be issue.

CS: Right. Well that's exactly what I believe. Let me ask you this. The theme of *Who Killed the Electric Car?* is that there is no real sincerity on the part of the OEMs to do this regardless of what they say. If they're doing this at all it's because they have to.

MD: Well, I can neither confirm nor deny the sincerity of the OEMs. Personally, I believe the EV-1 was a very sincere effort at GM, and that they would have never gone to those lengths if they had not been sincere about putting it into production. Was it successful? No. Did they handle the endgame of that program as well as they could have? Well, even they would not try and argue that. But you would never design a car from the ground up and engineer almost every part if you weren't trying to change a game somewhere.

This technology disruption looks different in the automotive industry than elsewhere. This is why I think we haven't seen a lot of success of classic technology disruption models working in the auto industry. Toyota put a great deal of effort behind the Prius, pushing it into production and selling two million cars. My hat's off to them. They changed the industry and they did it because I think they felt it would ultimately be in their long-term benefit.

What I'm seeing in the auto industry is companies that have lost faith in the future of stable gasoline prices. They believe very firmly that high gasoline prices, which they're certain to see once the economy picks up, will be drastically limiting to developing countries' automotive markets. So China and India aren't the gold mine that they need them to be unless they have a cheap energy source.

On the other hand, a cheap energy source is electricity. In the United States, we do have some regulation setting kind of a floor. I mean, we have the Zero Emission Vehicle Mandate, I can't remember if it's in 11 or 12 states now, and it sets kind of a floor that's in the neighborhood of tens of thousands of vehicles in the near term. But the scale that the auto companies need to make money is far greater than that. For what it's worth, I don't think we can determine, given all the management changes at GM, and earlier at Ford, what's really in the minds of the top executives. But I think we can say that the Volt, and I assume the LEAF, are going through a full automotive production development cycle. Those cars are being developed like they would develop their Buick.

If you were to tell me that they have no idea how this will all turn out, I could agree with that. I think they may have quite a bit of uncertainty as to how this is going to turn out. But I know how it's going to turn out. I mean, this is going to be the biggest thing since sliced bread and the only limiting factor will be can the costs come down to the point where it's profitable for the automakers. I'm not saying it's going to be their biggest profit; I don't think Prius is Toyota's biggest profit.

CS: No.

MD: At some point, the externalities are going to bear down on the auto industry and they're going to need to do something. I mean, the growth charts are probably a lot different for the United States than they were back in 2005, you know around 2003, 2004, 2005, the US was slated to use 60% more gasoline for the light-duty fleet in 2030 than in 2003. It's probably less than that now, but the point is that our population is still growing, we are driving more, every country whose affluence is growing is going to be driving more, and we are going to have to find real alternatives.

We can do some biofuels. That's a big advantage of being the United States, being the world's number one agricultural producer, and we have a lot of land relative to our population. It will get us part way there, but it will not get us to some sustainable future. And those realities cannot be ignored by the auto industry anymore. So they are going to have to do something substantive; I think they've felt that for a long time.

CS: By the way, let me ask you about hydrogen. People tend to think 1) that it's a "red herring, " as suggested in *Who Killed the Electric Car?* i.e., the OEMs love this because they know it will never happen, or 2), they believe Honda who says this is a legitimate contender for dealing with the range issue.

MD: I started off primarily a car guy doing automotive research at UC Davis and other places, and now that I'm at EPRI I've gradually become a little more of a utility person. The issue with that is that hydrogen has a huge infrastructure problem. That infrastructure problem does not go away. It's much greater than for electric vehicles. So the way for hydrogen to be successful is if the vehicles could be made at very low cost, to justify the high cost of the infrastructure. That's not likely to be happening in the near future.

CS: No, in fact, the precise opposite seems to be the case.

MD: Then once you get there, I think the issue also is that hydrogen requires a lot more energy than electricity per mile. So the simplest infrastructure would be sort of on-site electrolysis. That's going to use like four times the electricity per mile. So it uses four times the electricity per mile; you have really large equipment costs at the point. And you lose a lot of energy. So unless we're just generating enormous quantities of near zero carbon energy, you lose a lot of the benefit.

CS: Right.

MD: At the end of the day the problem with hydrogen is you need a near zero emitting source. You still have to compress it; that compression uses a significant amount of electricity by itself. It's just very tough. It uses a lot of energy. There are definitely niche applications in transit and heavy-duty and other things that you can use it for, but would those be enough to really sustain an industry and get your economies of scale? I don't know. I feel for hydrogen because I've been there. I was in the boat when they tossed electric vehicles overboard. So I know what it feels like when the technological focus and the policy focus shifts from one technology to the other.

In this world, a new automotive technology has about, I want to say, a seven-year window. It could be five years, it could be eight or nine, but you have like seven years to make something happen. Take hybrids; we're five years into this. They were getting to the point where they had to make something happen when we finally got the new Prius and the Ford Escape and the Highlander and the newer Honda hybrids, I mean they were five years into their window when we finally got cars that the average American wanted to drive. And then Toyota pushed this thing. Toyota with big contributions from Honda and Ford and you have the others doing their part as well. Being interested pushed hybrids into sustainable business where, granted probably not everybody is making money on it, but it's certainly on track and it's going forward. We'll see more hybrids and that's fantastic.

So the issue is, you've got this seven-year window. Electric vehicles had it. Hydrogen has it. Plug-in hybrids and electric vehicles for a second time have it. And I think that day one was probably January of 2007—maybe midyear. They announced the Volt, but it was midyear before people started believing it was real and other car companies started piling on with their programs, so maybe we've got until— Obama's million vehicles by 2015 is probably somewhat accidental—but it really would signal planting the flag for these technologies. You get to those million vehicles by 2015 and you've probably got a sustainable plug-in vehicle industry moving forward.

So that's our goal. So are we going to get vehicles on the market? Yes. Are they going to be successful? That's what we have to do to make them successful.

CS: I've never been able to get excited about hybrids. All the kinetic energy to the rear wheels comes from gasoline—you just have a black box in a hybrid that does something that happens to do with electricity. If you don't have a plug, all you're doing is changing the way you burn gasoline. Why didn't Toyota do this in 2004 or 2005?

MD: In 2004, no one was talking about this except EPRI and some utilities. There was one OEM program working with DaimlerChrysler on a plug-in hybrid sprinter van. I mean there was nothing.

CS: But I'm amazed that EVs had completely fallen off everyone's radar screen. In the late 1990s we had the EV-1 and the Toyota RAV4; the concept of an electric vehicle wasn't foreign to anybody. And everyone knew that represented the capability to move us away from oil. Everybody recognized that there is an issue with emissions, whether you're worried about global warming or whether you're worried about lung disease. So my point is, that this isn't new news. The automotive industry could have done this earlier. Why didn't they?

MD: Yes they could. But they didn't. Toyota could have. I built many plug-in hybrid vehicle prototypes with nickel metal hydride batteries, but I think the production folks sincerely felt that it was a steep enough challenge just to do a hybrid. I think that was the issue.

CS: Okay. Well that's cool.

MD: Yes, I mean, we could play hindsight all you want. What if GM had kept making 1000 EV1's a year just to study how they work—kept incrementally advancing the technology? Auto companies don't seem to do that, at least not in public. They tend to go after something like they are going to do it now and then if it doesn't work they move on and they do something else.

So whenever people ask me what EPRI's role has been, or who was responsible for all this, I would say "Did UC Davis show the world that you could build a plug-in hybrid that would do everything a normal car did with high-performance electric drive components? Yes, I think they did." And what did EPRI and the utilities do? Did we invent it? No. We incubated it. So, we kept the concept alive and moving forward and showed that it was not only technically possible but of societal value. We showed that it could be done and we kept it moving in that time between 2000 and 2007 when there was literally nothing going on. And so now that these things are emerging, we are trying to figure out the best way to plug them into the grid.

And I like where we are, because we can just plug them into the grid. That has always worked with new electrical appliances and it will work here too.

CS: Yes, exactly, unplug your toaster and plug in your car. I like it. What about smart grid technologies?

MD: We can plug these into the existing grid. Utilities will serve the load. It's the way it's always been. It's always worked. The industries always figured out how to make that work. If we can fully integrate them into the smart grid and utilities can work with their customers, who are the vehicle owners, to charge at the most beneficial time of the day for the person and for the system, we can maximize the benefits that the vehicles bring and minimize the impacts.

Now, ratepayers in general have to pay for a lot of the impacts, so we want to keep the impacts low and the benefits high. And a lot of that has to do with simple messages communicated back and forth between vehicles and the grid that convey owner preferences and requirements and system costs and the system conditions. So if you want to charge with the lowest cost electricity, the best way to do that is to feed the vehicle a 24-hour cost curve for electricity when it's plugged-in and have it cross-reference that against its owner's preferences and figure out when it's best to charge. And repeating this with tens of millions of vehicles will require very simple, very effective interfaces between the cars and the grid.

That means every vehicle has got to come out of the box ready to talk to anything it finds on the other side of the grid. Which is, hopefully, directly to utilities smart meters or to home networks, the devices that people will be using in increasing numbers to manage their own energy consumption. Every vehicle has got to be capable of doing that, it's got to have small low-cost communication technology onboard to be able to do that and be a living breathing part of the smart grid.

CS: What about V2G (vehicle to grid)?

MD: I see a lot of different possibilities for vehicle to grid. I think it goes without saying that people will want to charge their car at different times of the day, and that the smart folks who run the grid will figure out how to use that storage, incentivize people to be able to use that storage. So if you're at a utility that has a lot of wind power, that wind power tends to be really high at night and you'll figure out how to convince some of your drivers to charge late at night and provide a source for that power, a load for that power to go to.

CS: Will this be more about back-up power, wave-form stabilization, or both?

MD: Will cars be able to provide backup power to homes and buildings? Well there is some evidence that with hybrids, people have already rigged hybrids to do this in emergencies. There are a couple people that have weathered snowstorm outages with a giant inverter wired to the 12 V battery of their Prius. So the high-voltage battery would feed this 12 V inverter and when the high-voltage battery got low, the engine would come on for a few minutes, charge everything back up, turn off again—rinse and repeat for three days.

So can they provide backup power? Yes. Can they interact with a premise to help peak shave and energy demand? I think so. Will there be regulation services and peak shaving and spending reserves into the grid? I think that's to be determined. I think we need better business cases and we need to know more about how these would work and we need to know more about the overlap of other new technologies. For example, if utilities go after bulk storage to do things like aggregate wind resources were to help with ramping or integration of renewables or other system demands, those things will ultimately end up competing with vehicles in the V-G space. Because I always define vehicle to grid as providing contractual services to the system operator. So that's voltage and frequency regulation, up—down, spinning reserves, things like that.

I think we need to understand those business cases a lot better: the costs to the vehicle, the costs the operators incur, how that will compete with a big stationary battery or a compressed air energy storage plant or a pump hydro plant. We need to learn a lot more about these things, and how the grid is designed for power to flow to the source.

CS: Thanks so much, Mark. This has been great.

For more information on this contributor, please visit:
http://2greenenergy.com/renewable-energy-facts-fantasies/.

LOBBYING FOR ELECTRIC TRANSPORTATION

Readers of the blog at 2GreenEnergy.com know that I generally find corporate lobbying offensive, and that I believe that, in a perfect world, the entire practice would be completely abolished. Yet, out of fairness, and out of my respect for EV advocate Brian Wynne, I wanted to include a chapter on the subject.

Brian is president of the Electric Drive Transportation Association in Washington DC. As a result of several meetings, I have come to believe that the EDTA does a good and fair-minded job in promoting the EV cause. He and I happened to be in attendance at the "Business of Plugging In" show in Detroit in late 2009, where we sat down for this interview between sessions.

Craig Shields: Thanks for being here. May I ask you to just introduce the EDTA and its overall purpose?

Brian Wynne: Sure. EDTA is celebrating its 20th anniversary this year. We're a membership-based, not-for-profit, organization headquartered in Washington, and our primary function is to advocate for, i.e. lobby for, federal policies. Our members are inter-industry; we have members from the vehicle-manufacturing sector, we have utility companies and other energy providers, and we have component manufacturers such as battery makers. And we also have some of the newer business model folks like Better Place and GridPoint in the Smart Grid area, Coulomb in the charger area though those guys are providing infrastructure per se. In addition, there are some municipal participants like California Energy Commission of Kings County—some of the leaders in moving the electric drive toward the mainstream or electrifying transportation.

CS: So it's a lobby organization? In other words, it's a voice on Capitol Hill to make sure that policy is created that works for all its members?

BW: Yes, but first and foremost our mission is to make sure that policy doesn't disadvantage the movement toward electrifying transportation—and, if anything, helps to accelerate this technology toward the mainstream by doing several things. First, we support research and development. There's a lot of federal dollars that are targeted toward research and development. We make sure that this technology, whether it be battery technology or hybridization or component technology or fuel cell technologies—that we get our fair share of that.

CS: Well, if you're still here after 20 years, I can only assume you've had some success.

BW: Market incentives, like tax credits for hybrid vehicles—that was one of our wins. I'm referring to the tax credits for plug-in electric drive vehicles; we played a key role in crafting that. We actually crafted the language that resulted in the tax credits that are now available.

We've had a lot to do with demonstration and deployment money, and then most recently money either in loan format, or direct grants for battery and component manufacturing. So, all of these are different pieces to the puzzle. It all revolves around how public policy can support electrified transportation.

CS: Great. Let me ask you a little bit about each. Let's start with the last one first. Are we talking about stimulus money?

BW: Yes. That was the big opportunity: to get money from the government directed specifically toward advanced technology vehicles. Green jobs... those are the two words that everybody is using in Washington right now. How do we make certain that we are creating jobs for the future? That played right into our wheelhouse. The monies that we were talking about went up by an order of magnitude as people started talking.

CS: Yes, there's a lot of money. What have you gotten allocated so far? I guess if you include renewable energy more largely, you get a different figure than electric transportation more narrowly.

BW: Yes. For electric transportation more narrowly, you're talking about $2.5 million. So that's not an insignificant amount of money. And in many respects the timing couldn't have been better. Here we are, trying to make a transition to advanced technology vehicle manufacturing at a time when no one can find capital for the automobile industry. That's starting to change now, but for a while there it was difficult for anyone to get capitalized, and the auto industry of course was under a lot of pressure because of reduced sales and other issues, just generally speaking—they're trying to reduce capacity. So the stimulus money really came along at an excellent time.

CS: Good. Let me ask you about demonstration and deployment— what does that actually mean?

BW: Here's a good example that will illustrate this. When we started talking about stimulus funds, the utility companies said, "We want to see how this is going to work on our grid," and the automobile companies said, "Here's the configuration of the vehicle that we are planning on bringing out, we'd like to see how it works on your grid." There you have a project where there were the two of them that can come together with public support on a 50/50 basis. So the utility companies and the vehicle manufacturers were coming to the government saying, "We are going to put up 50% of the dollars, you're going to put up 50% of the dollars, and we are going to get a significant number of vehicles out there—like 100 to 300 vehicles—up to 1000 vehicles that plug into the grid."

CS: Just to measure and then extrapolate what this would mean if there were 1000 times that many?

BW: There's going to be a time of learning that goes on around those demonstration and deployment programs. There are two targets: one is to see how the consumers react. How do consumers actually drive these vehicles? How do they leverage the technology? Secondly, you have fleet operators. How are fleet operators going to use these vehicles? And of course, fleet operators use different metrics than your typical consumer does. For instance, I think in terms of miles per gallon, where fleet operators think in terms of cost per mile. So those are two different targets, but both are going to be required, ultimately, to get to the mainstream.

CS: Do you mind expanding on this thing about the tax credits?

BW: Think about the adoption of any technology. In the initial phase, you've got early adopters that are enthusiastic; they are excited by things that don't motivate the vast majority of consumers or fleet operators. Whereas the Prius has now become the second best-selling car in Japan, initially it was of interest to a very small number of people. Economics apply here—you have the cost of ramping up, the capitalization of new manufacturing, the technology itself particularly with regard to energy storage, lithium-ion batteries are expensive. How do we get down that cost curve? Well the answer is volume.

How do you get across Geoffrey Moore's chasm—you know, the Valley of Death? You need to reduce the price premium for the fleet operator or the consumer, and you need to change their metric, i.e., how they are thinking. The forecasting models are heavily dependent upon the price of gas and the price of batteries. We have chosen not to control the price of gas, so we need to work on the other side of that. The vast majority of the premium associated with the advanced technology vehicle is the cost of the batteries. So we crafted language, which Senator Obama, when he was a senator, along with Senators Hatch from

Utah and Cantwell from Washington State, introduced, saying, "The larger your battery pack in these vehicles, the greater the public good because the more oil you're going to displace, the more greenhouse gas emissions you're going to reduce, therefore the greater the tax credit." It tops out at $7500 for light duty vehicles.

CS: That carry 30 to 40 kilowatt-hours or something?

BW: No, actually, the top end of the scale is either 15 or 16 kWh. Now we're working on medium and heavy-duty tax credits. We had to give some of that ground back. There were kickers in the initial legislation for larger gross vehicle weights, and those got taken back in the stimulus package. We now have legislation in the hopper with some good champions that will help the medium- and heavy-duty side.

CS: Can you speak to the tax credits at the state level? Obviously there are 50 different discussions, but can you just summarize that?

BW: Obviously, some states are more activist in this regard than others—and we don't play at that level right now. But there are not just tax credits; there are other non-monetary incentives like HOV lane or carpool lane access and things like that, which are very valuable to particular individuals, obviously those that would benefit from having access to those lanes and so forth. So, there have been some states that have gotten out ahead on this and said, for one reason or another, "We believe this is really important for us." In some instances it might be municipal authorities that are looking at air quality issues and so forth. They want to incentivize people to use cleaner vehicles.

But if you want to look at states, look at Texas. Here's a state which is largely associated with fossil fuels, historically, and yet they've done an amazing job building out wind generation and are in the process of building distribution for that. And with people like Roger Duncan, who is on my Board of Directors, they are doing the nitty-gritty stuff to build this grid that will support electrified transportation. It's an amazing story.

CS: Well, to me, discussion of the way this breaks down at the state government level is very interesting from a legal and philosophic level. Take California as an example, where we are driving out businesses with extremely aggressive legislation on diesel. It causes an exodus of jobs moving to Arizona and Nevada, and it happens fairly rapidly. Arizona can say, "We don't have strict air quality standards." This makes a huge problem for all of us who are trying to make California actually work, where there is no level of consistency and fairness.

On a new subject, I think most readers have some level of cynicism about what happens inside the Beltway in terms of not wanting to "see how sausage is made." What can you tell readers who might be cynical about what happens in the back rooms of government and the lobbyists who deal with them?

BW: It's perhaps not the typical story that most people would think about. I'm working on a 25-year career at this point, including my work for a United States senator. The vast majority of folks that we are dealing with on Capitol Hill understand that, for reasons of energy security, for reasons of environmental quality, and for economic reasons, we have to embrace advanced technology for transportation. They understand that we need a grid which is more reliable, more efficient, and cleaner. These things are not really in debate right now.

But how you get from here to there, of course, is very tricky— particularly regarding renewables. Because energy is not so much partisan as it is regional.

That's always been the story in Washington regarding energy. Everybody's in favor of more efficient vehicles; there's nobody that's not in favor of that anymore. Even the automobile industry has essentially engaged in this battle: "How do we get to a completely different paradigm in a way that the consumer is actually going to benefit as quickly as possible?"

In fact, the vast majority of members of Congress have been coming to us and saying, "How can we help?" Now that's unusual in Washington; usually it's the other way around, where industries are going to Capitol

Hill trying to defeat regulations or trying to get the government out of their businesses. We don't play defense; we play offense. But now there are lots and lots of champions that want to be on this team, they want to get it done, they want to do the right thing.

The tricky part is not now; it's going to be two or three years down the road when members of Congress want to be able to point to results. Washington is a tactical environment driven by election cycles rather than a strategic environment, driven by the greater good. Now, I know that any good politician is going to say, "I'm doing this in the interest of the greater good, this is my vision of what's good for the country, etc." But suffice it to say that reasonable men disagree about that. Even reasonable politicians disagree about that. So there is a lot of tactical maneuvering that goes on. And that's not a bad thing. I don't cast aspersions on that.

CS: Well, most people I think would. Most people look forward to a day when there is broad campaign finance reform, driving more enlightenment and less parochialism.

BW: Well, we're going in the wrong direction. I wanted to think that when the Obama administration came in, that might have marked a turning point. The jury is out on that tactically speaking until after the midterm election—and if we have sort of a Newt Gingrich revolution on our hands at the midterm, like Clinton witnessed, I don't know where it goes from there. Essentially we will have been rewarding scare tactics and partisanship rather than what I sensed Senator Obama ran on, candidate Obama ran on, and what he came to office wanting to bring to Washington.

It is difficult to get people to change their tactical approach. In order to be successful, we have got to defend consistent policies. We have got to have consistent policies going forward because it's going to take a long time to transition a huge fleet. It's not going to work for every single vehicle model or vehicle platform. There are certain vehicle platforms today where electrification makes a dramatic amount of sense—economically, environmentally, commercially, however you want to draw it up.

CS: For example, small, intra-urban commuter vehicles?

BW: Better example: off road vehicles. There is a huge number of off-road vehicles that are completely electric. You look at the low hanging fruit, you've got port applications. You've got ground support at the airport—things like that. These are areas that we can influence directly with public policy.

But everybody's looking at the prize, which is the consumer, ultimately the vast majority in the volume of consumer vehicles or fleet vehicles, etc.; it's going to take a long time to turn that fleet over. How do we accomplish that in a tactical environment? How do we maintain the strategic vision and persistence, consistency of policy, that's going to be required to get this done? This is where our political system is going to work against us. I don't want to say it's going to fail us, because I don't know. We need to assess the cost of oil dependence, and see if we have some good politicians that can point to that and say, "The price of this dependence is too high".

CS: Let me just ask you about the cost of dependence.

BW: If you have a business with a certain vulnerability, you get an insurance policy. We are 97% dependent upon one global fungible commodity for the people and goods in this country. That is unacceptable. So if a politician wants to be associated with success, he wants to hold that up and point to that. We all want to be associated with success, so I'm not making politicians wrong for that. But they want to be able to say, "We deployed this number of vehicles inside this election cycle." But that's not going to happen when it takes 18 months for the battery manufacturer to get a piece of capital equipment functioning that's worth tens of millions of dollars, coming from Japan, which is the only place where they make them. We just don't spin up automobile grade battery manufacturing in this country overnight.

Our challenge at this point is to articulate in a way that politicians can be champions of a consistent policy over the medium and the long term. They can actually see this is making a difference and are moving toward it. It's like a front-end loaded mortgage. You pay a little bit now, you bring that deployment schedule in a little bit, and the benefits on the back end are enormous—the benefits in terms of oil displaced and greenhouse gas emissions reduced. But those benefits are further out in the future. It takes a lot of volume up here to get to those benefits. We just need to keep working on it, and be consistent about it. That's going to be the challenge inside of the Washington environment—and Washington is playing a critical role and will continue to play a critical role.

CS: Let's go back to your vision as to the role that both the public and private sectors should be playing to drive this forward.

BW: Well, it cuts in two different directions. From a policy perspective, we've got to craft policies that are going to benefit the greater good over the long term. That's not always easy to do when you've got a whole bunch of different crosscutting interests that are in that game. And yes, there are politics behind alternative fuels and who gets how much money, etc. The car companies themselves are not of a single mind on what the best way to configure this technology is. In fact, in many respects, they are all taking a slightly different approach, looking at different parts of the marketplace.

But to be fair, at this particular point in the adoption cycle, we are not entirely sure what the category killer is here yet. What's the iPhone out there for the car? Is there one? Does that even make sense? Does that even apply? The beauty is that electric drive is so flexible as a technology that it can be configured a lot of different ways. At some point we are going to need to standardize around certain things, but you can't rush that process. And, for crying out loud, you don't want the government trying to do that. The government will not make those decisions efficiently or effectively and the government has convinced itself of this, and certainly convinced business people of this, over the decades.

I think the government's helping GM in the bailout package was a function of people's jobs being at stake, and a large segment of our manufacturing economy being at risk. I don't think it was a statement that government wants to be making these decisions. The tricky part here is, and this is a little bit controversial what I'm going to say: if you were choosing a quick path to profitability for any automobile company, you wouldn't choose this one.

CS: You mean EVs generally.

BW: Right, EVs generally, because it's a new technology and it, ipso facto, takes longer to get to profitability with a new technology. The automobile industry in the United States, well worldwide, this is not a US phenomenon that we are dealing with—are doing this for different reasons and we are doing this on a different timescale.

I happen to be most focused on the US, but three weeks ago I was in China listening to the Chinese through interpreters saying many of the same things and asking each other many of the same questions that you're asking me today and that I get from other folks in the media. You know, "How long is this going to take?" and "What's the quickest path to the prize here?"

The answer is that's really difficult to know, and anybody who really thinks that they know hasn't studied technology convergence in other segments. And I've done a lot of that. I'm relatively new to the automobile arena, and I'm only here because of the fact that so many different technologies come into play here—and we play into the larger transportation system. But if you look at what's happening from a macro-economic standpoint, the price of energy is going to continue to go up; it almost has to; there is nothing out there that says you can plan on cheaper energy in the future. So if you take that as a given, we have to learn how to optimize our use of energy better. Transportation is a key portion of that. Geopolitically—I don't even have to go there. We understand that our money is now flowing; we are borrowing it from one place in the world and we are giving it to another place. This is not good for the world and it's not good for us. We are creating huge imbalances here that are not sustainable from a geopolitical or a macro economic point of view. We are doing this because we have to do this.

CS: So you are saying that the world auto industry is doing this but because it has to. In other words, that it's come to the realization that the current trajectory of this thing with fossil fuels, etc., is not sustainable?

BW: It can't possibly be sustainable. It's a limited resource. So the challenge we run into here is that I work in an environment that doesn't think in multiple-decade planning. But now we have to. If we look at the risks associated with climate change and we look at the pure economics of the price of energy going up because fossil fuels are a limited commodity and they exist in particular parts of the world... perhaps I'm getting too general.

CS: No, actually, I completely agree. By the way, for what it's worth, I think that you are right about the short-term. You're right about the next eight or ten years. I think you're wrong about the next 30 years.

BW: In terms of price of energy? You may be right.

CS: And that's simply because...

BW: We'll get more efficient?

CS: We'll see wholesale replacement of certain fuels, e.g., coal, with technologies, e.g., solar thermal. Whatever you say about renewables needs to include the notion that the price will continuously fall. It's about technology. Technology only gets better. Where peak oil means that oil gets pricier, technology always improves. In other words, we get more transistors on a chip, we get more Watts per square meter of PV, etc. now than we did two years ago. Renewables really boils down to one question: with what efficiency can you harvest the energy from the sun?

BW: Well, I can come up with probably a dozen examples of where you are right—but we still have gotten on an unsustainable path here with transportation. Unsustainable from the standpoint of safety, of convenience. You know, we clog our roads so badly in major metropolitan areas. It's unsustainable from an environmental standpoint if you believe half of what scientists are saying about climate change. And then there are the macro economic, the geopolitical elements of this. For all of these reasons, we have to change.

The problem is, unlike cell phones or WiFi or barcodes or, pick a whole bunch of different examples of things that have been enabled and are now indispensable because of the Internet. Because of what we talked about in our earlier discussion, this is going to take longer. People are really attached to their cars. I am really attached to my car. I'll drive my car till the wheels come off of it. I'm trained as an economist, and so buying a new car is painful for me. I like leveraging the investment that I've got in my existing vehicle and putting my money into something else.

I just got back from China, and had dinner with my good friend who runs an institute over there. He and his wife both drive vehicles like mine. That's what every Chinese is now moving toward. The Chinese went from making 250,000...I don't know how many electric bikes they were making a year, but now there are 800 million electric bikes on the road—that was inside of four years, and that was without government assistance. All of those people are now going to want to step up into something.

If we don't deal effectively with our traffic issues, we are all going to be riding a bus or we'll be riding a train because nothing is going to move in traffic. Now, if the Chinese can do with vehicles and liquid fuels what they did with cell phones—where they just leaped right over the terrestrial system and went right to wireless—we've got a prayer; they might actually show us how to do it. I think we need to pay attention to that.

The other element about Washington that I wanted to bring in here, we tend to oversimplify everything in Washington. We want to make it Fisher-Price. We tend to want to say this can be done easily, and as a result the expectations get out of whack and then politicians turn. They go from being champions to being haters—and we can't afford that. So, part of what we are trying to do in EDTA is be realistic, be credible, and say, "Yes, there is a huge opportunity here, and there are tremendous payoffs down the road, but let's not underestimate the challenges associated in getting from here to there."

Let's make a decision that we are going to stick with this—that this really is a medium-term thing. This isn't like the adoption of WiFi or something like that where the prices go from $4000 for an access point down to $65 at RadioShack. It's going to take a long time to get electric drive under the hood of the mainstream number of vehicles that are already out on the street driving around, let alone plugging into the grid in ways that are economically, commercially, environmentally sensible.

CS: We just heard a presentation from GM that, in my estimation, contained a lot of platitudinous garbage. But mixed in were some ideas that I thought were on track, e.g., the metaphor about the flat screen TV and the prices falling down. And it is true that a company like General Motors has to pay the price for the fact that we just don't have the scale to bring down the price of EVs; somebody has to pay for that.

BW: Right. The mainstream consumer is using metrics that are essentially only appropriate to an older transportation system. Even though we think we've made some good investments in electric drive, at what point do we start migrating the kind of monies that are being invested into protecting and sustaining our existing fuel system into this new fuel system? That gets inside Washington's internecine warfare, and you will have big battles between the oil industry and the other players.

CS: I am asking you to comment on certain things that may be sensitive, and I'll understand if you don't want to. Here's one: if you are a vice president of anything at GM, or Ford, or Toyota, the value of the 401(k) is really a function of your ability to sell and maintain profitable relationships with customers that include cars, parts, and service—then a new car every few years. Here you have EVs that go 500,000 miles before they even need a tune-up. So if you're an OEM, you may be doing it because you have to, but you certainly aren't doing it because you want to. Do you consider that true or not?

BW: This is an industry-wide movement. It has to be. We are talking about people working, we're talking about money being made, we are talking about the next generation of jobs. We are talking about technologies being leveraged here that don't necessarily originate in the automobile business. We are talking about the grid, and the information technology and back-office systems; there are so many different moving parts here. So, the answer to your question is I think it is good for them. I think it's got to be good for them, and the reason I say that is because this is blood sport.

The automobile business is a low-margin blood sport business. But how did Ferrari survive? Well, they survived because they've got a differentiation. That's not low-margin; these guys differentiate around technology in many respects. That's not how all of them make their money, but there are really a couple of crosscutting things here, whereas at the end of the day yes, some of these companies their business models over the past have been based on high-margin trucks and things like that. The Prius wouldn't have been as successful, the Prius wouldn't have been possible, without faster microprocessor speeds and better software. That had nothing to do with the automobile. You know, that started in a completely different place. Now vehicles are so computerized, and why shouldn't they be? That's how we create optimization.

So if I am working for any automobile company right now, I am recognizing that advanced technology is the wave of the future and I want to be differentiating around advanced technology. That's the only point I was making.

CS: And that may trump what I was saying, though personally I don't think so. I think that if they could have turned the clock back to the day that they were selling Oldsmobiles to retirees every 2 1/2 years, they would have those days back in the snap of a finger.

BW: And these guys are not monolithic in their thinking. There are tremendous debates going on inside these companies. I have been privileged to be party to several close-up, and there's phenomenal skepticism within these companies about advanced technology and how it ultimately lands in the marketplace, and how you get down the cost curve. There is even greater skepticism in the utility industry about "why do we want to confuse ourselves with this? Isn't this a bit of a distraction?"

CS: Well, where I was going with this thing... even if we can agree that big auto is ambivalent, big oil is not at all ambivalent. I think it's clear that Big Oil wants to kill this. I moderated a panel at the Alt-Car Expo in Santa Monica last fall where I pointed out that Chevron's asking us to imagine them, an oil company, as being a part of the solution—that is simply too hard for some of us to imagine.

BW: Well, that's painting with a pretty broad brush. A lot of people would say things like that about politicians, but there are some people like Jay Ainsley, I shouldn't single one particular individual out, but if you read his book *A Flawless Fire,* this is a guy who puts all these pieces together. He's a smart guy who really gets it. He should be commended for that. And he's from a place where it's not an accident. People in that area vote for people who care about their environment. We just need more of that. We need more people that are going to connect those dots.

CS: Thanks so much, Brian. Great speaking with you.

For more information on this contributor, please visit: http://2greenenergy.com/renewable-energy-facts-fantasies/.

RENEWABLES AND CLEAN TRANSPORTATION AT THE MUNICIPAL LEVEL

Over the past half-century or so, the public sector has played a critically important role in the development of technologies that have changed our lives for the better. Despite problems that we all see with our government here in the US, we need to count on our elected leaders at the federal, state, and local level to act on our behalf, and to encourage the eco-friendliness. Rick Sikes, manages the vehicle fleet for the City of Santa Monica—a municipality that has taken quite a progressive stance with respect to environmental stewardship. Rick agreed to speak with me and show me around his facility.

Craig Shields: It's a small world, Rick. You may not be aware of this, but we were both panelists at the Alt Car Expo in Santa Monica last fall. I moderated the discussion on EV charging infrastructure—I think you were in that auditorium with your group immediately before I showed up.

Rick Sikes: Yes, I saw that.

CS: I note that you folks in Santa Monica are extremely progressive in terms of what you've done with alternate fuel vehicles. Do you want to tell me a little bit about that?

RS: Sure. Back in the probably late 1970's early 1980's the oil embargo from OPEC hit—there were actually two incidents way back then—everybody was affected by the same things, lines at the gas stations, and so forth. So it was decided that we shouldn't be buying all our oil from one place, and we should reduce the amount of oil we use. And at the same time there were a lot of the environmental regulations coming into effect for cleaning the air from the AQMD (Air Quality Management District) and CARB (California Air Resources Board).

And you're right—Santa Monica is progressive, and wanted to do their part. So they started looking at alternative fuel vehicles and REFP—Reduced Emissions Fuels Policy. They started buying some propane as a lot of people did, and natural gas, ethanol and whatever else was out there. In 1984 we had the first city Reduced Emissions Fuels Policy, and then about 1998, the city began a sustainability plan. So that was another driver for using alternative fuels and reducing our use of petroleum.

CS: Okay. So and as we sit here this afternoon 87% of the your fleet is AFVs (Alternative Fueled Vehicles)?

RS: Yes, 87% of the municipal fleet. So the way it breaks down is we have everything except the Big Blue Bus (local bus system) and the fire department. So that 87% includes the police, the carpenters, plumbers, the water department, the solid waste collections—all the administrative vehicles or whatever else is used to run the city, the bulldozers and loaders and dump trucks all that kind of stuff. Now we pull out of that number the vehicles that are exempted by the state of California law as police emergency response vehicles. So those are mostly all Crown Victorias; the newer ones are capable of running on ethanol. We don't have ethanol available to us in the city so they're still on unleaded gas. But the pickups that the parking enforcement, and whatever else for police use, and so forth, are alternative fuelled.

CS: That's great. Well tell me if you would about electric versus other alternative fuels. What have you done with respect to EVs?

RS: Well, we've always been involved in EVs since the 1990s when they first were mandated by CARB. So we've had EV-1; we've had the Nissan, Hondas—whatever was out there, the Rangers, etc. The last batch of vehicles that we were able to keep electric were the RAV 4's.

CS: Toyota?

RS: Yes. And you know about the cars being pulled off the market including the EV-1 and stuff like that. Plug-in America and its predecessor DontCrush.org was active in trying to save those vehicles; they asked us to support their efforts with Toyota, which still had some vehicles out on the street and they were starting to pull them. So we wrote a letter to Toyota asking to keep the electric vehicles—and let us get some more—because we'd already had experience with electric vehicles, we didn't require them to do the maintenance or to support us. So they let us keep them and get some more—so right now we have 24 leased RAV4's and one that we purchased.

And then we have a number of neighborhood electric vehicles, including GEMs, which were the first ones on the road—and also some Miles pickup trucks and we've got 80 total now.

CS: I would think that the neighborhood electric vehicles would be perfect for parking enforcement. How's the quality?

RS: There are maintenance differences and issues with everything. That's why I have a job. So I wouldn't say that there are quality issues; we have trash trucks that are about a quarter million dollars and that's where we spend most of our time. You'd think that if you spent that much money on something you'd never have to touch it, but that's not the case.

We did have a parking vehicle built for us. We don't have delivery of it yet but we had one designed and built specifically for our use.

CS: That's not a bad idea. Whom did you hire?

RS: We went to an electric car company in Long Beach called "Good Earth." We went through quite an effort trying to find a parking vehicle that would be electric. We said, "I'll give you a PO—when do I get delivery?" And they said, "We don't know—we'll get back to you."

CS: Makes it hard to buy something when you don't know when...

RS: When you don't know when you can actually buy it.

CS: Right. Well, obviously the whole electric transportation industry is facing that to one degree or another. Unless you want to stand in line for a Tesla or something like that, you can't buy one of these things even if you wanted one.

RS: Can't buy one for love or money.

CS: I think everybody would like to believe that this situation of supply of quality available electric vehicles is going to burst open in the next year or so—the LEAF, etc. Where do you see this going in terms of your job?

RS: Well I think we're a little further out than that. Because I don't have a vehicle in my hands, I would say that we're two years out to get any number of quality vehicles that are available.

CS: What is the gating issue on something like this? Can't these things be manufactured and sold profitably? I can't imagine the problem with getting capitalization to start cranking them out. What is the issue in your estimation?

RS: I think part of the problem is you have the large OEMs that have traditionally supplied vehicles sell 10 or even 15 million vehicles a year; at an average price of $20,000, those are huge numbers. Manufacturers are focused on the product that is driving those revenues. And they need very good research and quality control before they produce limited models. They're not going to bring anything out unless it's complete.

CS: For sure. Moreover, the OEMs in my estimation have incentive to not do this. If you're Ford or GM, there's nothing good about them at all.

RS: They have no reason at all.

So then you go down to the lower end of the small manufacturers like neighborhood electric vehicles. But the numbers are tiny; two years ago it was about 4,000 neighborhood electric vehicles were sold a year, and a little better than half of that was GEMs. So you go into all the remaining neighborhood electric vehicle manufacturers, split up 2,000 vehicles...

that's pretty small numbers in each.

You could say that these small companies have the maneuverability because of their size and just one purchase order, like we give, can attract investment capital for them. So then they can start producing it and afford to be able to put together some manufacturing process and sell them at some kind of a reasonable price that they can make a profit off of.

But there aren't that many opportunities like that and that's one of the roles that I think is important for government at all levels, the federal, state, and local. It's nice that we're able to do it, but it's disappointing that a local level is going to drive the success or failure of a vehicle.

CS: You're saying that the public sector could break this kind of vicious cycle, this chicken-and-egg thing.

RS: It has to; that's the only way to do it. There's no incentive for business to do it because it costs more money, and they're looking at the bottom line. So unless they're mandated through carbon rules or carbon credits, there's no reason that a private company would go out and say "OK, I'll disrupt my operation and get people to use a piece of equipment they're not trained in, something that's new and unproven and costs more."

CS: Right, there's no upside there.

RS: … Unless it's good PR or the incentive to embrace sustainability or environmental issues. So I think it's the responsibility of government to help drive that. Nobody else will do it. Philanthropists only go so far. And even the good will of corporations will only go so far. They'll run up against a wall at some point in terms of return on investment.

With the financial crisis we have this next-quarter profit—this short-term thinking. What will be the return for our shareholders next quarter? And if we cut these jobs and we do certain things, that makes money. But if you want something good to happen down the road ten years, that's where the planning and responsibility through government comes from, because we're not looking for a profit.

CS: Right. I don't want to sound nasty, but I am of the opinion that Corporate America doesn't do things philanthropically except when they're visible, like the Ronald McDonald House. Maybe I'm missing something.

RS: No, it's true. I used to work for UPS and it's a great company; I love it and they have a very big commitment to environmental programs. They don't pollute where they don't have to—they outfitted their jets before the laws came into effect for fuel efficiency and noise and things like that. They're generally ahead of the curve. And they look down the road but they still, especially after they went public, they shortened up the viewpoint. And with GM and all the other companies, it changes by the quarter. And you can't survive in the long run like that.

CS: And by your wits, does the city of Santa Monica have that long-range vision?

RS: Yes, in fact, they've simplified it. They bring it down to a policy of sustainability, so that basically means that we're not going to use resources wastefully or unwisely. The only word anybody ever has to remember is sustainability.

CS: That's great. But then why Santa Monica and not Dubuque? I'm not trying to get you to criticize Dubuque—I just picked them out of the air. But why is Santa Monica unusual if not unique in terms of its commitment to sustainability?

RS: There are a number of factors. The elected officials are committed to the concept of sustainability—as is almost all of the staff in the entire city—as are most of the residents of the city.

It is a big reason that people come to live here. You get disagreements all the time but by and large, the people who live here are proud of the fact that we have the beautiful ocean and we try to keep the air clean and we try to take care of the environment. To be honest, there's selfishness in it; we make money from it—our biggest industry is tourism so we bring people here and we want it to stay like this. You get the same mindset with most of the people here.

We're also the affluent population so…

CS: So you can afford to do things.

RS: Yes. We have that desire but we also have the resources. A lot of cities are stuck without the resources. They're cutting back everything. We're able to look at a long-term goal and say, "Now this is worthwhile," even if it will cost some money up-front.

CS: Can you give me a good example?

RS: Sure. We put a natural gas station in and a couple of natural gas vehicles early on—but the real commitment happened about 1996— when natural gas cost more than diesel or gasoline. Two years ago when you were reaching $4.00 a gallon on gas and over that for diesel. We were paying $1.25. So if you hang into it long enough you'll see the results.

CS: Is this something of a political issue? I mean, would you say that locals tend to be liberals, and tend to be more interested in sustainability? I heard yesterday that 79% of college-educated Democrats are concerned about the global warming theory—but only 19% of college educated Republicans are. That's an incredible number.

RS: Yes, when you get past the Orange Curtain (south of Los Angeles into conservative Orange County), environmentalism becomes a bad word. Here it's a good word. It's funny, you get insulated. I've only been with the city for six years, but there are a lot of people that just don't get the concept that there are places where you go that if you talk about environmentalism it's like cussing—it's a bad word.

And you can see that now in some advertisements for even cars and stuff. They're not selling it on the environment or emissions; it's "efficiency." That's the new buzzword is "efficient," not "sustainable."

CS: So as to avoid the politics?

RS: Right. It's not an environmental agenda.

It's interesting—whether you believe in global warming or not, it's been proven that along transit corridors that health is much worse. That we know that we're giving money to people that want to kill us.

CS: Exactly. How controversial is that? I know they're people who don't believe in global climate change and so forth. But are there people who don't believe in cancer?

RS: Right. There are so many reasons to get rid of oil; to me, it doesn't matter which one you choose. We suck it out of the ground and burn it and we'll have a problem where we open a void and the water table is polluted. Just look at the way it's manufactured. It's pulled from the ground and it destroys the earth and you ship it around and make clouds of smoke.

Regardless of global warming, just take that out of the picture all together. I live over here in the marina and I get this black soot on everything. It just coats stuff. If you were out in Bakersfield, you'd get dirt. If you're someplace else you might get snow. Here, it's black soot.

CS: That's terrible…. Please tell me about the Alternative Car Expo.

RS: It was started because there are conferences and exhibits for industry and for different specific fuels; there's the Electric Vehicle Symposium and the Electric Drive Transportation Association conferences, and there's a natural gas conference. So this includes all of the different alternative fuels. It's really geared toward industry, specific to vendors, putting them together with fleet operators and getting the vehicles on the road, because that's where most of this stuff happens—in the fleets first. It's a proving ground.

So this was put out there to help educate the general public that this stuff's here—you don't have to wait. The first year we did it at the same time as the LA Auto Show—on purpose—because it was the alternative.

CS: That was a bold move. If I were backing it with my own money I'm not so sure I'd do that.

RS: Well the next year we didn't, because we ran into vendors who said, "You know, we didn't think we'd be playing with the OEMs that much," and as it turns out we do have OEM involvement. But the OEMs have a problem with trying to do two exhibits and two staffs and all of that stuff. But it was especially the smaller companies that just can't do it. You know they can barely do what they can do. So to try to do two shows at the same time is just impossible. So we changed it. And the focus is alternative transportation and energy.

CS: Great. Let's go back to EVs for a minute, if we can. I'm obviously disappointed that companies like Phoenix and other companies with apparently good concepts bit the dust.

RS: It's all in the timing; they get out there too early or don't get enough sales. We had a couple of Phoenix EVs on purchase order and they're still sitting there for whomever will take them over. My personal opinion with Phoenix is they tried to do too much with it; they tried to make it longer range than it had to be—and getting into fast charging and I don't think that's necessary.

I think that's exactly what is holding back a lot of people from electric vehicles—this misconception of how they think they drive today versus what they really do and what they really need. Being that we're a small city we're able to do some things with electric vehicles—like we have a medium-duty electric service truck that'll be delivered in the next couple of weeks from Zero Truck. It's limited to a 75-mile range but when I wrote the specs with them I told them I don't need anything more than 50, because we can't go 50 miles in a day.

If you look at the way people drive, 80% of the people go less than 30 miles a day. So that's a good chunk of people. Our South Bay Cities Council is putting together a study over four years—80+ % of the trips that were generated went less than 4 miles. So people going down to the school or to the grocery store or whatever...80% of the trips.

CS: Well it's 80% of the trips, but not 80% of the miles.

RS: Well, certainly not 80% of the miles. But if you do that much local stuff and then only occasionally go farther, you can buy a natural gas car. People who want to go visit grandma could rent a car or take mass transit.

There's this idea that your one vehicle has to do everything. Once in a while I go down to Baja. We're a two-car family. I don't need both vehicles to be able to go to Baja; I just need one.

CS: Personally, I think that as soon as we have something like the LEAF that hits next fall...attractive, inexpensive—I think these things will sell like hot cakes. Obviously they'll sell well in a progressive city like Santa Monica, but I think you'll be surprised at a very quick adoption curve. Where do you stand on this?

RS: I hope you're right. I think that there will have to be some driving force behind it. I don't see people adopting it unless fuel prices goes up or we tax energy or give incentives. There've been incentives out there for natural gas vehicles, and Honda's been out there for 10 years. It works; there are no problems with it, but I don't see that it has been adopted as much as it should be. You can get natural gas anywhere in Southern California, you get natural gas at home—you can put a fueler in your home. The car goes 300 miles on a tank of gas. And it's neutral cost to purchase it because it's about a $4,000 federal tax credit, and then the fuel is less.

CS: I see. You know, to my shame I couldn't tell you anything about this. I lived in Woodland Hills for 15 years, and I couldn't tell you where there's a natural gas station there.

RS: Right. That's my point. People don't see them. They're out there, but you have to plan your trips. You couldn't buy one and just forget about it. There's a station a block away from here. You have to pay attention to what you're doing and people don't want to; it's too much trouble.

It would be good to start incremental change. People will convert over to a series hybrid like the Volt, where they can operate in an electric mode only, for a period of time and then, if they have that comfort to get rid of range anxiety. But now we're into 2015 before there are enough of them out there that people have used them and say, "Now I can replace it and I don't have to have this range extender on it," and get a straight EV.

One of the future things I think that will happen, and I didn't believe it as much before, is I think that we will go more toward hydrogen in the future. It seems that people want one fuel, whatever it is, everybody wants to pick a winner.

CS: And you think hydrogen could be it?

RS: To me it doesn't matter. I've seen propane, diesel, CNG, electric and stuff, so it doesn't really matter to me. But a lot of people want to know what fuel they'll be using. You can't use electric for everything.

CS: Well you can't use electric for everything right now, in the absence of battery swapping or fast-charging.

RS: If you got an 80,000 GVW truck going cross-country, you can't run it on batteries—not with the technology that we have now or in the foreseeable future. But you could do it with a hydrogen fuel cell. We have a project that we're negotiating right now for some hydrogen fuel cell trash trucks. So they're just electric trucks and instead of using a battery pack they use a hydrogen fuel cell. So you can make hydrogen from bio-diesel or any waste.

CS: I think you're only the second person I've ever met who agrees with Steve Ellis (see interview with Honda's Steve Ellis) on this.

RS: I was thinking that we would have regional fuels; that's how I thought this thing would play out eventually. Iowa would have ethanol, somebody would have natural gas, urban areas would only use electric. I don't think the battery swap will happen.

CS: Not in the United States, at least. If you're Israel, I can understand it.

RS: And maybe China could mandate it. We have too many different cars here. If you buy a Nissan LEAF, or a GM, Ford, or Toyota—and they all decided they would play the same game and have exactly the same battery pack, that was connected and attached exactly the same way and have exactly the same power out of it, it would be possible. But none of those things will ever happen.

CS: They certainly haven't happened in the past.

RS: Eventually they'll do some fast-charging. I don't get hung up on waiting for fast charging. We don't need range. The city of Santa Monica adopted a resolution to ask the state of California to allow medium speed vehicles which are NEVs "plus." That would be the quickest, easiest way to promote and grow electric vehicles.

CS: That's interesting. Will that happen, do you think?

RS: It's happened in nine states already, contrary to federal law. So something's gotta give. Either somebody's going to die and the Feds will shut em down and hang somebody out to dry and kill it, or more likely, with the pressure from the state of California and other people, it moves forward. We need green jobs, we need to do something about transportation and energy, and we're changing our car culture and thinking, so we have an opportunity right now that we could make it happen.

CS: This means 35 MPH?

RS: Yes, the one we're pushing is 35 MPH maximum on 35-MPH streets. If you look at it realistically, the accidents that you have in an urban environment are either intersections or pedestrian bicycle type fatalities...and those would be reduced. You would presume if somebody were coming up to a red light and they have a 6,000-pound SUV and then punch it and now all of a sudden they're going 60 when they hit that intersection. If you're going 35, you're gonna stop.

There are a couple of things that can be done under the same laws. They've done NEV lanes in a community in order to connect cores.

CS: Fascinating stuff. Great speaking with you.

RS: You bet.

For more information on this contributor, please visit:
http://2greenenergy.com/renewable-energy-facts-fantasies/.

INTELLIGENT ENERGY
MANAGEMENT IN BUILDINGS

As a clean energy legend Bill Paul says, "The best way to clean energy is not to use it in the first place." And, this, of course, gets us into the subject of efficiency. Up until recently, I've viewed energy efficiency as similar to dieting. Sure you can consume less, but you do so at the expense of some level of deprivation. At a certain level, renewable energy offers the hope that we can consume energy like utter pigs—perhaps driving a 600 horsepower Hummer or heating our swimming pools in February.

There are a number of interesting discussions that issue from this. But they're all kind of moot insofar as we're a long way from abundant and inexpensive clean energy. Until we get there, we need to concern ourselves with efficiency. Let's dive into that discussion in this interview with Echelon's Steve Nguyen, in which we explore the subject of using information technology to drive up the energy efficiency of the world's buildings.

Craig Shields: Most of this book is about generating renewable energy—and I think that's exciting. But until we get to the point where we have abundant clean energy, we have to be careful with the utilization of the energy we have. I suppose that's the premise of your business, right?

Steve Nguyen: Yes indeed.

CS: Good. We've been hearing about intelligent building management for decades. I have a friend who told me in the late 1980s, I think, that I just HAD to buy your stock. And I know I would have made a killing had I taken his advice. So where are we in this overall evolution of this exciting industry?

SN: We are at a point where most commercial buildings, meaning office buildings, go up with some form of automation system—typically confined to islands of automation, at least in the North American market. There is some form of control system in the heating, venting, and air conditioning—some form of control system in the lighting, some form of control in security and access.

Now when I say "islands of automation" what it really means is that many of these systems are disparate; they function as silos—but they tend to be managed by the same teams within the building. When we talk about making the building intelligent, we tend to take these disparate systems and combine them at the user interface level. In the "large building" sector, the focus is on heating, venting, and air conditioning which is one subsystem and lighting which is a second subsystem. And those two systems in that market generally constitute somewhere around 70% of the energy load.

So commercial buildings typically have that level of control. Below that, you have very limited functionality on the control.

CS: Like a thermostat that somebody pushes up or down?

SN: Exactly. Typically what you'd find in those buildings is a thermostat on each floor—usually behind some kind of locked box—that controls typically the whole floor but sometimes sections within the floor, no light harvesting—that's it.

CS: Could you define "light harvesting" for us please?

SN: Light harvesting is a function where you take into account how much ambient light is provided to the building and then you dim the lights in those areas accordingly. So the perimeter area in a building that has a steel spine with a glass shell facing north, south, east, west— the walls are going to have a lot of glass—and those are typically going to be either offices or common hallways. There's no reason to light them during the middle of the day. There's a lighting controller that's combined with either a timer or a light sensor that dictates that those lights should be off, so someone doesn't have to actually turn them off. And that's the minimal sort of light harvesting. An example of something that would be more aggressive would be the Gare de Nor, the North Station in Paris where they have light tunnels in which the light is guided through reflective panels and tubes, reflected off of a mirror, which literally harvests the light. The light comes in and then it gets distributed; that's a very advanced form of light harvesting. It still includes sensors but it also includes servos to guide the mirrors, and other things of that nature.

CS: Wow. Well maybe we should take a step back and look at where we were 50 years ago. We had nothing like that, right? We had somebody who came in the morning and turned on the lights and cranked up the heat.

SN: Yes. A few decades ago, control was done through air pressure, through pneumatics. About 20 years ago the controls norm had become something they called a direct digital control. So something like a centralized thermostat or a centralized lighting panel and things like that.

CS: How does pneumatics work in this context? I'm not following that.

SN: Well, air pressure was used to control the dampers. Instead of an electronic control, you had high or low—a squared off sine wave. And that would be it, just on or off basically. The air pressure is literally the signal to change a state of a damper; it's a pulse of air going through a tube. That's really old school stuff.

CS: OK, I'm with you. Echelon is essentially a protocol right? A protocol which is to buildings like TCP/IP is to the Internet—so all this stuff can work together?

SN: Yes. Let me give you my short explanation. We created a control protocol and platform, aimed at individual devices working together to perform some sort of function. We figured out how to take silos of functionality and integrate them into a single system.

Our approach was to say, when it comes to control, everything is the same—whether it's an aircraft, a building, a home, a tunnel, a parking lot or whatever. When you strip away all the nuances of "What am I doing with the control," the signal is essentially the same; that's really the epiphany we had. What does that mean in terms of a control signal? If you're closing a door, the protocol that we invented sends a signal that says "I'm open" or "I'm closed." That's it. If you're closing your garage door, if you're closing a ferry door, or if you're closing the door at a ski resort that signal is the same. If you're sending weight, it's the same, it's in kilograms—and that's the same whether you're weighing a ship at a shipyard or weighing chemicals at a pharmaceutical plant.

Now our customers then take that technology and they build systems and products. In the building automation stage—that means people like, Honeywell Home and Building, Johnson Controls, Siemens, Phillips, Hubbell Automation, Samsung Heavy Industries and so forth—they've adopted the technology upon which to build their products. What they add is all of the features and functionality on top of that.

CS: Do you mind speaking as best you can to the highest and best of building control systems that are built on top of Echelon? In other words, not so much what you provide but what they provide. For example, how exactly does one use this to conserve energy in the industrial and commercial building space?

SN: Then I'll do that, using our own building, which I'm most familiar with. We have a very intelligent building with an integrated control system. That means that every system in our building is on a single control network—that includes automated blinds, lighting, fresh air, conditioned air—so heating and cooling, security access, the elevator, outdoor sprinkling, parking lots, exterior lights and so on and so forth. All of those systems are on our technology. But when it comes to energy conservation, we focus on two, as I said before lighting and HVAC.

CS: That's just because, that's where the lion's share of it is, anyway?

SN: Absolutely. On average, it's about 70%. In our building it is almost exactly 70%, it turns out. Despite the fact that we have hardware labs in our buildings, since we do circuit design, 70% of our electricity is consumed in heating, venting and air conditioning and lighting. Now, as individual tenants within the building, each of us determines, as an individual office, how we will conserve electricity. So in my case, I determined that whatever my light's level is, in the event that energy needs to be conserved, it will drop to 20% of that level. Each of us has an office and so we can each determine how our office reacts during an energy event.

CS: An energy event like a brown-out?

SN: It's a little bit subtler than that. Pacific Gas and Electric determines typically within a 24-hour notice that they need a specific percentage of energy off the grid. It's called a demand event. They work with energy service companies to say, "Can you effectively guarantee, that when I need that 10% or 5% of electricity, you will reduce need by 5 or 10%?" And so they have contacts with a number of companies to help them do that. They could do it directly with us as an individual building owner, but more often than not, there are service companies that help do that. They can aggregate all this energy. That's a type of insurance program.

CS: Okay I understand. So you said then that you will do what exactly?

SN: So I say, as an individual, first of all that I will allow my office to participate, and then I determine the behavior. So in my case, my behavior is: if my light levels are above 20%, they will drop to 20% of max. Whatever the heat level is, I will allow my office temperature to go up four degrees, cause the demand event will typically be in the summer in California. And that office next to me may be different. They may decide, "Well I'm always hot anyway, so I'm not going to let it go up more than one degree, because I can't work, and I need brighter light because I do different work than Steve so I have brighter lights."

So for us as a corporation, our individual employees determine how the offices react. The facility management company then determines how the common areas all act, things like the hallways and the unoccupied meeting rooms and the lobby. Then our response is set up into three separate tiers of responsiveness. This means that PG&E will tell us how severe their need is and publish that need 24 hours in advance; they typically will know that far in advance. The automation system in our building subscribes to a web service from our energy service provider. That web service typically will only say date and time, duration, and severity. Based on those little bits of information, our whole network will respond. We as individuals don't do anything beyond that first setting. So, in a matter of three minutes, we drop about 30% at a max, of our energy consumption.

CS: Amazing. You're somehow compensated for that, I presume.

SN: Exactly so. There's an incentive in the form of either direct credit or a lower contracted energy rate. The energy service provider takes a percentage of those things. So the benefit to us is we don't have to do anything actively. Previous to that, we would actually broadcast a message and employees would then do something based on the behavior they choose to implement, and the control operator would log in and manage the reload areas and do this all kind of manually. Then we would have to then come back and prove to PG&E (which is easy to do with a control system) that we actually did what we said we would do.

We don't have to do any of that anymore. We just sign on as a customer, and in our case we're signed on to a company called Enernoc. They have the contract with PG&E and we respond and we get a benefit.

CS: I'm with you. Let's go back to these discrete, disparate silos of information. Normally, that's a bad thing; people want to integrate these things. But it's unclear to me how that's possible here. For instance, how much light you need to your job at 2 PM and how hot you care if the room gets, are fairly independent as, in terms of variables.

SN: They are. What becomes important is the idea of the control network itself. So the granularity of the response determines the long-term participation of the end user. It's like a diet in this case. The more drastic and draconian the requirements, the less likely you are to continue with the protocol. So in our case, because we've embedded the intelligence into what our customers do, they've used our technology to embed intelligence so the tenants really don't see the activity. They don't necessarily feel the activity, because it can be very gradual, and it can be automatic.

For example, if there's an energy event and you cut off the air conditioning, you get warm and if you also cut off the fans, because you're not measuring air pressure, you get stifling. And then all of a sudden everybody's super-uncomfortable. And how does that manifest itself? Here's a funny example: the first time we did this, it got hot and stuffy and the president of our company decided to go work in the gym. So, here's our highest-value, highest-compensated employee deciding that he's going to the gym for two hours instead of working. That's a big deal. So if you translate that to everybody in the company being hot and uncomfortable that's not a good thing, you saved a lot of energy but your productivity is down significantly.

By being able to control everything in the environment we can avoid that. It can be very selective and very tailored. It begins to cross all the boundaries because the systems are integrated; they can work together on very little data, and we can program the entire system to work cooperatively.

CS: That's wonderful. I presume this all has an upper limit of what can be achieved, right? Won't we get to a point at which we've done everything we can about the instantaneous management of energy within a building, vis-à-vis the human beings that live there?

SN: It has more of an asymptotic trajectory to your point. And that's true; we ourselves are reaching that asymptotic benefit, because we've been energy auditing our building for ten years on a quarterly basis, so our incremental change is lower than anybody else's.

CS: That's my point—and yes, asymptotic is the word I was looking for. It's not like space travel, to take a weird example. If mankind is still here in 200 years, we'll be farther and farther into space with more and more knowledge of the universe around us. I don't think you could say that about your industry though I may be wrong.

SN: No, I think that's pretty accurate in the sense that we will reach a point of diminishing returns. But we are so far from that point. In most parts of the world the control system consists of a switch in the basement, and these buildings hemorrhage energy. They are single walls with non-UV coated, single pane windows, they have no control systems whatsoever. And that is probably 90% of the world's buildings.

We have a whole market left to conquer.

Commercial buildings are making progress, but 90% of the buildings are not in that space. Look at the average franchise operation, the average bank branch, the average convenience store—they don't have controls.

We worked with a bank that wanted to do centralized energy management and control. This is maybe five years ago that this project started. When they went and audited their 1000 locations, we found that 70% of them ran their air conditioning or heating 24 hours a day. And 30%—it might've been 40%—had their lights on 24 hours a day. So, just in the discovery phase they saved a massive amount of money. And that's before they started putting out the CD-ROM that told the local maintenance guy how to hook up what they had to the Internet.

CS: Fantastic. This has been very, very valuable and I'll certainly send you a copy of the book.

SN: Thanks!

ACTIVISM AND THE MEDIA IN CLEAN ENERGY

For 14 years, Rona Fried has worked hard to tell an important story: there are numerous, potentially profitable businesses and investment opportunities built around sustainable living practices. Her website, SustainableBusiness.com boasts over 100,000 unique visitors per month— each coming to learn more about this exciting subject. When I visited Rona near her home in Huntington, NY earlier in 2010, we developed an instant friendship that has blossomed into a partnership between Sustainable Business and 2GreenEnergy.

Craig Shields: Why don't we just start with your website Sustainable Business—one of the most well developed and well respected in the world. What makes it unique?

Rona Fried: The fact that we started it in 1996. It was one of the first websites to focus on sustainable business in the world. We did start it at the dawn of the Internet. At that time, the state of the industry was completely different than it is now. It was still a fairly small, fragmented community of people spread around the world that were interested in this. What was most useful about using the Internet at that time was its ability to find people all over the planet who were not connecting with each other and bringing them together. That was why I started SustainableBusiness.com—to do all the things that I am doing now.

Then, just as now, we were providing information—whether it was news about the advancement of sustainable business or jobs or letting people meet one another through business connections and all those various things. No one was doing any of that at the time. So in order to find a job, for example, there were printed newsletters that you could subscribe to that came out once a quarter. They would have some pretty general jobs that ran from biologist to scientist to engineer to various kinds of what I would call "old-line environmental" jobs. Of course, a lot of them would be already filled by the time you received the newsletter, so it was just a completely different way of providing that kind of information; there was nobody at all providing news on a daily basis.

Our focus at that time was all on positive news. So the idea was to let people know about the positive things that were happening—what companies were doing to advance sustainability within business. That emphasis shifted over the years as there became much more news and at some point it just didn't become realistic anymore to just provide positive news. It would look silly at this point to have a news feed that focused just on what companies are doing in a positive way, because there are now thousands of companies that are doing things—where at that time there were dozens.

The whole field has morphed completely. There just wasn't all that much news back then; we basically covered everything that was going on. Now of course you could never do that. So a lot of it has to do with how the field has just multiplied tremendously in leaps and bounds in the last 14 years. I could not call it greenbusiness.com because using the word "green" was considered hokey. So, all of these things have changed over the years. So that is what made us unique, i.e., that we were the first to pull all of that together and provide news and jobs and ways for people to connect with each other on an ongoing basis.

CS: Right. Well, speaking of green, where do you see this going from the standpoint of the perception of the word "green"? I have interviewed certain people who have suggested that we have hit a wall, that people are saturated with this idea. Do you consider that this is going to be a movement that represents a huge paradigm shift where in a few years everybody will have some level of eco-sensitivity, or do you think it has hit something of a wall?

RF: I think that is really hard to know. I do not really focus on one word versus another word. Having said that, there was a period of time probably about five years ago where people tried not to use the word "green" because it turned off so many people and so they focused on high performance buildings or advanced buildings. Now I see, people feel completely comfortable using the word "green" and they use it all over the place. So, I think "green" has gained an acceptance, I think that a lot of the negative connotation associated with it has gone away. There will always be naysayers; there are global warming deniers too.

CS: That is for sure. I find it intensely difficult to try to keep up with what is happening in all these areas at any depth. You have a staff, but I presume there are challenges associated with keeping your finger on the pulse of what is happening around the world on a day to day or a week to week basis. Can you speak to that?

RF: It has become an extremely noisy universe. It has really gotten to the point where you can look at it like any other field, as if you were an expert in technology or health care or any other sector of our economy. Are you going to be able to keep up with every single thing that goes on everyday? Absolutely not. Our field has become like that—and that's a good thing. What is hard for an Internet business is that it has become incredibly noisy in terms of the number of websites, the number of blogs, the number of competing—and redundant—information services.

CS: Exactly. There are sites that are simple news-feeds that just do not even aspire to original content. And that must make that a little bit difficult.

RF: It has all become very difficult, especially in trying to make a living doing it. We compete with countless free services doing similar things—they may not be doing it at the level of quality that we are doing it at—in fact I can tell you for sure that they are not—but that does not matter to a lot of people. It is harder to be found and it is harder to stand out. Where as a few years ago it was still easy. I am happy to say we are still at the top of the pack here but now the difficulty is trying to get through all the obfuscation as opposed to standing out for what we do.

CS: What do you see as the frontier in terms of sustainable living and eco-stewardship? What are the most important principles that people overlook?

RF: I think an area that is really hard to get on the big radar screen the way renewable energy has is sustainable agriculture. It is a huge industry, and it is enormously destructive for the environment.

CS: Are you talking about slash and burn agriculture in the rainforests, or are you talking about Monsanto and things like that?

RF: I'm talking about Monsanto and Cargill and ADM and big agriculture that produces our food.

CS: Could you take a couple minutes and talk about the most egregious, unsustainable practices that are common in agribusiness?

RF: Well it has been shown now that agriculture contributes somewhere around 30% of global warming emissions. By producing our food organically rather than using chemically intensive agriculture we could make big dents in climate change. And so it is very important that agriculture goes through this transition. Just like the fossil fuel energy has to go through the transition to clean energy, agriculture has to go through the same transition from chemical based agriculture based on mono-cultures (massive single-crop farming) to organic agriculture based on improving the soil and based on using many crops and rotating them. Of course the use of genetically modified crops is a very difficult barrier to achieving organic agriculture.

The main idea that Monsanto and others have tried to put forward in their propaganda is based on lies. But they are so powerful and so diligent in putting forth their agenda that people believe them. Monsanto has said all along that one of the main benefits of using GMO (genetically modified organism) crops is that you do not have to till the soil, which is a lot of what destroys the soil, and destroys the microorganisms in the soil. They said that it would reduce pesticide use, because they inject these GMOs into crops and supposedly that means you don't have to use as many pesticides. But the opposite has proved true. Farmers that use GMO crops are actually using more pesticides.

Obviously there is a lot of controversy over the health effects of taking genetic material from a completely different organism and putting it into another plant, and we really don't know what the health effects are of that. Those GMOs can drift onto organic farms and spoil those crops. So it is all very important and just as difficult an area to tackle.

CS: What do you think should be the role of government in moving what is essentially the private sector: agribusiness, big energy, etc into more sustainable practices?

RF: The role of government has always been to provide funds for the basic research that is necessary to commercialize new industries. That is what the government did for the Internet—which is why we have the Internet—because in many situations the private sector is not going to pay for the basic research needed to develop an industry. So the only one to pay for that is the government. And it plays the crucial role there.

CS: What about incentive and creating a "level playing field?"

RF: What we are trying very hard to get the government to do right now, unsuccessfully, is to put a price on carbon. And that will unleash the innovation that we are all looking for to get us off and running to a green economy. As long as we are not including the so-called "externalities," all of the costs involved in producing fossil fuels, or fossil fuel based agriculture—whatever the industry—we are not including all the costs involved. In these cases, those fossil fuel based industries are always going to be cheaper and the green economy cannot emerge until we have this level playing field.

CS: That is exactly what I believe. So what you are saying is not so much about providing incentives but removing artificial incentives?

RF: Both have to happen. We have to switch; we have to move past this tipping point.

CS: I have always said that if you take away the subsidies that are poured into oil and coal and so forth, and you force the consumers and producers of fossil fuel based energy to pay the true and complete costs of these things, you will have renewable energy this afternoon. Those businesses will be gone in a heart-beat. Do you agree?

RF: Absolutely. Then there is also a role for regulations. We have got ridiculously high cancer rates and I think that if we finally get to the bottom of why we have such rising cancer rates, it is because of lack of regulation of such huge amounts of toxins that we have in the air and the water because companies are not regulated, chemicals are not regulated, toxins of all kinds are not regulated. And they need to be. We need to have a society that says "No. You cannot put poisons in the products that you make for people to use."

CS: Right. Now how do you get people to care about all of this? I mean obviously people care about cancer, but in a world where there are pressing worries, where people are losing their jobs, and turn on the television and they see houses on fire, cats stuck up trees, etc. With all the distractions in people's lives, why should people care about something that is essentially long term in its nature?

RF: Well there are two answers to that. One is that a lot of what is happening is not long term; it is happening right now. That is something that you can look at with our conversation about the rise in cancer rates. People need to make the connection between the products they are using in their everyday lives and the effect on their health. If you can look at every single product that people use and say "Gee, it is not that hard to understand that we are really having a negative impact on the health of society and of course all the species that we share the Earth with by using these toxic products."

The other piece is the role of government. Government is there to look out for the best interest of the people in general. And so it is really not up to people to be able to solve the climate change problem. But it is up to governments to take a higher perspective of all of this and understand that they need to create a platform and set the rules by which society operates, so that society can continue to operate. I understand that the average person does not connect the dots necessarily with climate change or if they do, they feel like it is totally out of their control. I think it is an absolute travesty that the government does not require cars that get the highest mileage possible, for example. People should not have a choice when they go to a car dealer to get a gas-guzzler; it simply shouldn't be an option.

CS: Well, here you are talking about a couple of things. First of all you are talking about an enormous amount of power, concentrated in the federal government. And I am not sure that most Americans are going to want that level of intrusion.

RF: Set some rules, that is all. I am just saying, we need the power to set some rules. The Obama administration finally raised the car efficiency standards, the CAFE standards to 35 miles/gallon by 2015— an increase that we could not get done since the 1970's. So all that had to be done was many years ago they needed to keep increasing those standards. And they did not, because of industry lobbying.

CS: I have come to know you as a very sincerely dedicated and committed person with respect to these issues and I am sure several others. You must have had, in the 14 years you have been doing this, some very rewarding experiences in terms of influencing people to do good things. You have over 100,000 unique visitors a month—that is a lot of traffic. Tell me about some of the most rewarding experiences that you have had.

RF: Well, seeing the field grow from just a pimple to something where you hear world leaders talking about the subject—they all understand what we are dealing with. To see renewable energy grow from nothing to being the industry of the future. To see green building grow from nothing to being completely mainstream and at the point where just about every single town and city is incorporating green building standards. To see smart growth take off. I mean basically every single concept that all of us have been working towards is coming to fruition. It has taken a really long time—and we have had 8 really terrible years under the Bush administration which took us backward instead of forward. It is absolutely happening and going to happen, I just pray that it happens in time.

CS: Yes. It's funny you mentioned that, that very last phrase there because that is what I was going to conclude with: Are you essentially optimistic or pessimistic about the way this is going to turn out?

RF: I am really on the fence. I am optimistic that everything that we all want to happen will happen, I am not optimistic that it will happen fast enough. I would have to say that I think that there is no getting around some climate change at this point. We are seeing it, I am seeing it out my window right this second. But I do think that we are going to have a very difficult period of time? I think we are going to be having a lot of very severe weather events that are going to be catastrophic. I think that we will get through them and that at some point things will shift extremely rapidly. I think that there will be entirely new industries that do not exist today, for example that remove the carbon from the atmosphere. I think it will all happen. I even think that at some point the world will understand that it cannot just allow people to produce endless children and that we will actually have to control our population. But will the transition to any of this be easy? I do not think so.

CS: OK. Anything else you'd like readers to know?

RF: Here's my take on the Obama administration versus the previous administration. I think that this just does not get talked about or is understood enough out there. It is like night and day. People when they look at the Obama administration tend to only look at what is going on superficially in terms of what the media is talking about, i.e., the major legislation that he is not being successful right now in putting forward. But what I look at is #1 his attitude, which is 100% opposed to the Bush administration attitude. They saw anything to do with the environment as a cost to be avoided. Obama really understands the importance of keeping our environment safe and healthy. He has made some excellent appointees to every single agency that deals with the environment including the USDA, Department of the Interior, as well as the Department of Energy and the other agencies that are involved. If you look at what each of those agencies are doing, there is really a sea change in their approach from the Bush administration.

I was so impressed listening to Secretary Salazar, Secretary of the Interior, and this is something that none of us would really hear about, but he is really looking at "conservation on a landscape scale," were the words that he used. This is hugely important because more and more we have a very fragmented conservation network, where our lands are protected. Animal species and plant species are having a very difficult time coping with climate change, coping with endless development on the part of human beings and invasive species that are now coming in and taking over, and killing millions and millions of acres of trees. Taking a landscape approach is very important to bridging the gaps and bringing these fragmented communities together. So my take on the Obama approach is that it is absolutely essential. He is doing a fantastic job as best he can with hands half tied.

There is no comparison; he has been undoing all the terrible things that the Bush administration did for eight years of unraveling and not enforcing any of our environmental laws, and of course continuing to subsidize the fossil fuel industry at the expense of the renewable energy industry, among many other things.

CS: Good stuff, Rona. Thank you so much.

THE ROLE OF THE MEDIA IN CLEAN ENERGY

Renewable Energy World is one of the largest websites associated with clean energy, boasting a staff large enough to cover this subject with astonishing breadth. Stephen Lacey spoke with me about his organization and its activities.

Stephen Lacey: You keep yourself busy; I see a lot of stuff coming out of 2GreenEnergy.

Craig Shields: Well, I enjoy what I do. You seem to be all over the place though. If there were about 20 or 30 of me, that's the productivity that I attribute to Steve Lacey.

SL: It's neat keeping up with what's happening in the industry; I just keep on learning about this fast-growing market. There's always something new going on and there's never a dull moment—so I just have a blast doing it.

CS: I'd like to hear, from your perspective, how this burgeoning industry spawned another kind of sub-industry of the people who are covering it. Please tell me what you see as the development of this industry from a media perspective. Maybe we could start with what you see as the central challenges?

SL: Well, from our perspective as an industry publication we're covering all the technologies and markets around the world. So it can be very difficult to keep our fingers on the pulse of everything that's happening. As I said, growth is occurring so quickly, we really need to be on the ball. So it's just the diversity of topics that make things fairly difficult.

We have a great staff, and we have a lot of fantastic freelance writers that have been covering environmental and renewable energies for quite a while, so we feel like we're on top of it—but it's an ever-changing industry.

And this is tough for the mainstream press. A lot of reporters don't come from an engineering background and don't understand energy markets. It's just from covering this industry for so long I now have a better understanding, but general-interest reporters are hopping from energy to health care to broader environmental issues. So a lot of journalists in this industry and a lot of business professionals conflict with some of the mainstream journalists that are covering the industry.

CS: Okay. Could you give me an example that illustrates that point about renewable energy world, vis-à-vis the mainstream press?

SL: If you look at the balance sheets of solar manufacturers, and you go out and you talk to a lot of the manufacturers out there today, you're seeing increased demand for product. While we've moved through a very difficult period of time, most people in the industry would say that the solar industry is picking back up. And I still see a lot of reports about how solar is far too expensive. We've seen radical price drops because of the lack of demand over the last six to nine months. But that demand is picking back up and I see a lot of journalists off the mark, not really seeing that demand pick back up.

There are so many good reporters out there covering these issues and the increase in awareness of renewable energies generally is very heartening to see. But I also do see a lot of misconceptions out there as well. Yet having said that, we're not perfect either.

CS: Okay. Well let me ask you this. You're the perfect person to answer this question.

I personally try to pick the winner with respect to renewables. If I were the king of the world, I think I would do the research and then pick whatever I favor a continent—say, North America. I'm not sure I would try to do solar thermal, PV, wind, wave, tidal, geothermal and so forth—all at the same time. Where do you stand personally on this?

SL: I disagree with you, and I think most people I talk to disagree as well. Given the variability and the distributed nature of the resources, we need more distributed variable power plants to capture that.

I am a big fan of local generation as well. I firmly believe in citizen empowerment in reducing infrastructure and that by implementing more distributed energy systems we make things less costly in certain cases. The range of technologies that we have out there is very exciting and our ability to mix and match those technologies based upon the site-specific needs is important—especially as the cost of these technologies comes down. A one-size-fits-all solution is really difficult to implement—just given the variability of the resource. You also have a lot of security problems, both from a power plant security and from a resource security perspective.

When we centralize these power plants, the cost and time it takes to build out new transmission lines and permit those transmission lines and integrate them into our already complicated transmission network.

We can avoid a lot of the regulatory and social friction problems with more distributed power.

CS: Okay. Well that makes sense. I'm sure you come across people who favor high voltage DC as a method to solve this transmission problem, and energy storage with molten salt—especially where the energy is in the form of heat already.

SL: Definitely. I hear proponents of centralized generation, saying let's get all of our resources from the southwest and ship those electrons all around the country. When I hear conversations about that it's more theoretical than practical. I think people say, "Hey, we can do this"—and this is theoretically possible; renewable energies can power, most if not all of our homes and offices and industrial buildings. The ability is there, I just don't know that everyone thinks it's totally practical.

CS: I know I'm certainly in the minority on this but that's fine. New question for you, Steve: Can you tell me your approach to "Web 2.0"—in other words, how you have driven a huge base of fans and paying subscribers?

SL: Absolutely. We started off realizing that to drive people to the website we needed a newsletter. And that was really going to keep people coming back to the website. So, we started up the site about 10 years ago, and we started our newsletter shortly thereafter—actually bought up another larger newsletter from a gentleman who was doing something in the New England region and accumulated a thousand people and then five thousand people and then fifteen thousand people. We realized that we needed a captive audience through email to develop that subscriber base that would keep coming back.

Our first success was realizing that it was all about the newsletter. That was also an important revenue generator for us as well. Because the people on those lists are people who are likely to buy products from companies, more likely to listen to company messages and so the companies can then advertise in those newsletters and do customized ads to our subscriber base. So that was a real driver of traffic for us.

Once we had that audience it was all about developing really good content in house and working with a lot of other writers who were already in the industry to put together content that people would want to just keep coming back to. It wasn't about regurgitating press releases; it was the commitment to do our best to put out as much comprehensive, original content as possible. By doing that we focused on quality over quantity—and people really liked the product.

As we've evolved in the last couple of years, we've really tried to empower our audience on the site. These are really standard tools that you see in most websites—just really good comment boards. Now we're starting to allow our users to create profiles on our website so that they can go and post information about themselves, network with other users, post blogs, and try to facilitate more interactions around the content that we're creating. We want the conversation to continue on our website; we don't want people to just come and read an article and walk off; we want people with similar interests to talk to one another about the topics that we're discussing—and to help us perform coverage as well.

We've found that our audience is so well educated—I'm always blown away by how smart people are on the website. And I get a lot of interesting angles from the stories that I cover just by reading the comment boards and hearing what others have to say. So audience empowerment is a big part of what we're trying to do right now. We're really trying to experiment with that. We get people clicking through on the site and we increase page views, which increases ad revenue obviously; it increases the quality of our site, so again people just keep coming back because they connect with people and they really want to hear what others are commenting on.

Again, these are all common tools that we take for granted these days—but there are a lot of renewable energy sites that just aren't doing this. There really aren't any at this point. There are some that have started to succeed and other social networking sites specifically for renewable energy that have not been doing fabulously. But this is a real need for our industry.

And in terms of search engine optimization, on Google and Yahoo, links are currency, so just having that really good content allows people to link back to it in their blogs. The more people that link to us, the better we'll show up in the Google rankings. It took us a while to move up the ranks but as the site got more and more popular, eventually we had that top rank that's really given us more and more traffic. And we've been around for about 10 years now.

CS: Yes—and that longevity obviously helps as well. But I'm so impressed with the functionality and the creativity—the bells and whistles. For instance, you do contests right? The best idea for sustainable building design—or whatever—right?

SL: Yes. This is the first time we're rolling that out. It's a renewable energy award for best project…most innovative CEO, and so forth; we have a variety of project categories. Again, this goes back to empowering our readers—we want to hear what they have to say about the coolest projects that are out there, the coolest companies, and new technologies. Because we think they have a lot to say.

We've gotten a great response from that, we've had hundreds and hundreds of submissions to the awards and then we'll have a conference where we'll be presenting the awards. We'll actually make it part of our 10th anniversary party as well. The more and more we can interact with our audience the better. That's really what it's all about. We'll have some editor's choices but we'll also be picking them from the most popular according to our audience.

Let me just ask this of you: 2GreenEnergy is focused on finding market opportunities for up and coming companies? Is that how you would describe what you do?

CS: That's certainly part of it. What the site is really about is providing information with respect to the science, the business and the politics of the migration of renewable energy. So there are information products, there are consulting and investment services—everything from writing or editing business plans to raising investment capital, to providing marketing services. "Do you have any research on X?" or "Can you get me funding for this?" or "Can you help me find distribution channels for my product?"…whatever. The answer is always yes.

SL: Well, please blog on the site; I think readers will respond well to you.

We merged with a larger publisher, Penwell Publishers out of Tulsa, Oklahoma about two years ago. And they've given us a lot of resources;

they've helped us out in a lot of areas. We've hooked up with some other print magazines and we're really happy with the relationship. But the group that runs the website is just a very light, nimble company over here so we pride ourselves on that.

CS: That's wonderful.

SL: Fabulous, thanks Craig.

CS: Wonderful thanks so much. Talk to you again soon.

For more information on this contributor, please visit:
http://2greenenergy.com/renewable-energy-facts-fantasies/.

RENEWABLE ENERGY— BUSINESS AND FINANCE

Renewable energy legend Bill Paul has recently agreed to write financial reports and newsletters for us at 2GreenEnergy. We're delighted to have come across such good fortune, as there really is no one more qualified for the task. Bill retired from the Wall Street Journal after a 20-year stint as a staff writer covering the energy sector. Highly respected as an analyst in this space, he's been interviewed on CNBC and numerous other television channels through the years. We're thrilled with the association. I thought readers might like a glimpse into how he thinks and where he sees the renewable energy industry going from the perspective of international corporate finance.

Craig Shields: If we could just maybe reprise the conversation that we had at lunch that day, in which you talked about all the fascinating aspects of the various world economies—and financial pressures and trends that have driven renewable energy and electric transportation into the spotlight.

Bill Paul: Energy is at the crux of all things financial. There is not a single business that does not have energy as a significant cost factor, there is not a single country that does not have energy as a significant national security factor, and there is not a single household that does not have energy as an important budgetary factor. With energy, you are operating on all levels in terms of finance, as well as national security, as well as health consequences. You have privacy issues, property rights issues, basically any personal choice issue on lifestyle and you have a perfect storm of political, economic, and social interests—all competing for "the right" approach.

CS: What are the privacy issues?

BP: When you talk about the smart grid, which is the next point in the evolution of the electric utility industry, you are digitizing the utility industry. This will permit marketers to access people's information and businesses' information, in terms of their energy use and consumption. A company in the future will know how much time I spend in the living room and the bedroom, and how many minutes per day my refrigerator door is open. It will be able to craft sales pitches aimed specifically at those patterns. That raises many serious questions that are equal to if not above the questions raised by privacy experts over the Internet.

CS: Well that is amazing, good point. This all seems so far advanced from where the utility industry has been in the past. Do financial advisors understand this stuff? What if I'm managing my 401K and I go to my broker and I say, "Bob, what do you think of the energy sector?" How likely is it that a randomly chosen broker is going to be on top of all this?

BP: Your average broker will, at best, have a handle on oil and natural gas—the fossil fuels. And that will only be because the traditional energy analysts at his brokerage are sending out reports on a regular basis. The fact of the matter is that, with rare exceptions, there are no real good analysts in the United States today who are concentrating on what you can call the new energy industry—which is essentially the move away from fossil fuels into more than two dozen different sources of alternative/green/clean energy information. Not the least of these is energy efficiency itself, the so-called "mega-fuel"—

which is probably the cheapest way to control energy costs and protect national security.

But going back to your question, brokers may pretend to know what they are talking about and they know a little bit about oil and gas, but ask them about renewable energy and they will hit a key on their computer and up will pop the solar companies that trade in the United States—and only those companies. Most brokerages do not even want to know about companies whose stocks trade outside the United States, because it is a little more difficult to trade those stocks; the float is a little lax in some instances. So what you are left with is an overindulgence in a handful of solar stocks, not all of which are really very good.

People are myopic. Hence they do not know about the dozens of other exciting companies with stocks traded around the world. Take geothermal, for example, with 15 or 20 publicly traded companies, most of which are outside the United States, some of which are fantastic in terms of their potential. Brokers are not going to know the first thing about ocean and wind power—or clean coal technologies that are, of course, another aspect of all of this.

CS: I would think that part of your job as an analyst in this thing is to not breathe your own exhaust. I would think that there is propaganda being spewed out from all manner of sources. One might be big oil and gas—trying to make everybody believe that there is no such thing as peak oil, but there is such a thing as clean coal. What is the truth on this thing and how do you get at it?

BP: Today, as we are speaking, it is virtually all about fossil fuels, in terms of their practical presence. You cannot run an automobile unless you have oil refined into gasoline. You cannot turn on your lights without coal that has been used to generate electricity. That is now. As you move forward in time, the reliance on fossil fuels inevitably must decrease. I say inevitably because it is axiomatic that fossil fuels are finite and simultaneously the world population is growing rapidly—as is people's never-ending desire for an increased standard of living. So whether or not there is peak oil in terms of the amount of oil physically that can be extracted from the Earth, we are approaching a point in time where the amount of oil that can be extracted is running dangerously close to the demand of that growing population.

So, you can say that peak oil is anywhere from 2013 to 2025; that is my reasonable time frame. It is unreasonable to adopt the fact peak oil is here today, which the environmental community often does, and it is unreasonable to adopt the time frame that peak oil is 2050. You have to put all of these factors into play so that they can work off of one another: increase in population, rising standard of living, and the doubling—I repeat, the doubling—of the number of cars on global highways between now and 2030. You simply have to come up with more energy. Don't think of it as saving the planet so much as bumping up against the limits of a finite source of fuel, namely the fossil fuels. So now you inevitably must add into the mix the so-called renewable fuels, fuels that essentially never disappear: wind, solar, geothermal, ocean waves, and so forth. You must move into these fuels to augment your existing supply.

By 2020, if we do not have a major change in the percentage of renewable versus fossil fuels that is in use, our standard of living as a nation and as a world will have collapsed; there is just no way around it. We do not have the amount of coal, much less the amount of oil or natural gas. There are some out there right now screaming and yelling about how much natural gas we have, forgetting the fact that those tight rock formations essentially will be drained quickly. That is just a fact of life; there is a lot of gas there today, but it is going to get used up quickly. You have to think out the time-line—and that time-line says renewables must be a significant percentage of everybody's energy consumption within ten to fifteen years.

To get there in fifteen years, we have to start now. In fact, we are already behind on the journey. Obama just announced yesterday the first two nuclear reactor units in this country, which if all goes right, always a big IF, will be online no earlier than 2017. We have only two base-load sources of electricity: nuclear and coal. We are not currently building any new coal fired power plants. Period. They cannot get permitted. An ugly little fact of life, that no broker is going to know, is that while we get 50% of our electricity today from coal, the number of coal fired power plants being shut down by their owners is rapidly on the rise. The number of plants that are now 40 years plus in age is rapidly rising. That is a very critical time in the life of a coal-fired power plant because the owner must decide whether to invest a fortune into life extension work or shut it down.

Any broker today should understand—but none of them do—what the options are for the world to get its sources of power and motor fuel in the future. Even in the best of times, Wall Street has maybe half a dozen analysts who really understand the electric power industry. I am not just saying the electric utility industry, but rather the electric power industry, because utilities are only one factor in that. More and more of manufacturing is becoming electrified, moving away from oil. The next step will be transportation, moving away from liquid gasoline over to electric fuel. And you must understand also that the electric power grid is being reinvented to a smart grid which will change how we live, work, and play. Analysts do not get this and I do not know if there is enough time, frankly, for Wall Street to get up to speed on this; there really are not enough guys. I know the guys who are out there, the best ones are all very long in the tooth now already, and I do not see any kids coming up who really understand it.

CS: So what you are saying is that there is a lot of activity that needs to take place today, and there is a lot of investment that needs to take place today.

BP: This crisis is probably five to seven years out. And the technologies to solve the crisis right now are just starting to be discovered. But realize this: that is the point at which they are most valuable to an investor. You figure out what the best technologies—the winners—are, right now, and the companies are in existence today. Microsoft, Google before it caught on, or the first guy who discovered oil in Western Pennsylvania—all of these breakthroughs had an early day and in those early days was when the big money could be made, because you could buy low. Right now today, there are literally hundreds of companies, more than a thousand actually, that are developing the next generation of energy technologies that will be at the core of harnessing all of these renewable power sources.

We are in the early stages for all the good stuff right now. And the investor who gets in today, and by today I mean from 2010 to 2012, is the investor who in 2020 will be very glad he made those decisions. Energy is the biggest industry in the world—and it is slowly but surely converging with the other biggest industries in this world: telecom, IT, and transportation. All of the major industries in the world are beginning to coalesce into the new energy generation.

Of course, some are going to fall by the wayside and die. Eventually, however, the ones with the really good technologies will, in my estimation, be acquired by the big boys because energy is such a big industry that only the big boys have deep enough pockets to play in this game. GE, Philips, Johnson Controls, IBM, Siemens, ABB, all of these monsters are going to be acquiring dozens of companies each to fulfill their specific needs. And you will see these technologies blossom into a multi-billion dollar enterprise. For anyone with a five-year time horizon, I do not see a better investment opportunity than new energy.

CS: I'm ashamed to admit it, but I tend to take a U.S. centric view of the world. And I know from talking to you that the most exciting possibilities you see are actually happening elsewhere. Can you speak to that please?

BP: No one will ever confuse the European Union with China. But in point of fact, both are passing government policy measures that are propelling the alternative energy industry forward in ways that the United States is not; we are getting so far behind. For example, by 2020 in the UK, every residence will have a smart meter, bar none—so the smart grid essentially will be in place to a great extent by 2020 in the UK. You have motor vehicle rules being passed, you have biofuel rules being passed by the European Union, you have biomass rules being passed in the UK, you have waste-to-energy power plants being passed, you have carbon recovery projects, taking the CO_2 from a coal-fired power plant, capturing it and transporting it to the North Sea where it will be used to pull more oil out of the ground. All of this is happening essentially under EU edict—and there are hundreds of companies that are in the business of meeting these new requirements. Meanwhile, halfway around the world, you have the Chinese government that is simply doing all of this unilaterally. It is a battle today between the EU and China in my mind, with Korea and Japan separately trying to play catch up and the U.S. twiddling its thumbs and yelling from one party to the other. There is a conspiracy theory out there that China did its best to muck up Copenhagen to keep the US political cauldron boiling over on climate change so it could, very quietly, keep pushing full throttle on development of all this stuff. In China right now they are building wind farms so fast that half of the generation capacity cannot even be hooked up yet to the power grid.

They'll be putting in high voltage power lines—ironically some of the very best, most efficient new technologies being used in power lines are coming from a couple of companies in the United States who cannot make a single sale in the United States but they are making a small fortune in China.

CS: Wow. What's happening in Nepal?

BP: Ahhh...you may have gotten me there.

CS: I was kidding. I just picked an obscure country.

BP: Nothing to my knowledge in Nepal, but there is some very cool stuff going on with wind in Mongolia. The Chinese have branched out there because the winds are very, very good. Just like there used to be gold prospecting, now there is wind prospecting; there is ocean wave prospecting; they were all over the Tokyo Bay about a year ago looking for the best current. And again, each one of these things requires the technology to capture and convert it into usable electricity.

Now, what is the beauty of that? We use less coal; it has less environmental damage. I think it is the IEA's (International Energy Agency) numbers that say we have a requirement for 50% more electricity globally by 2030 than today, with at least 20 to 25% increase just here in the United States. That is an enormous amount of increased demand. Now think about that and realize that there is not a single coal-fired power plant being permitted. Now you see why renewables have no where to go but up.

Let me hasten to add that this all comes a cropper, if the government does not get its act together on transmission lines. We need, right now, something like 1,000 miles more of new high voltage transmission lines. And we need them to be smarter than today's lines because they are going to need to be able to incorporate into the mix all of these intermittent renewable power sources.

And here's another thing: no one has done the hard preliminary work on ensuring reliability of the electrical power system in this country. As usual, the government has gone off, half-assed, ordering up all of these new wind farms and solar projects—and absolutely nobody knows with any certainty how they will effect the grid—and whether the lights will in fact stay on. I mean that is just remarkably stupid.

Now you couple that with the ever present travel of state regulators, each of whom has a little fiefdom that can hold up a necessary power line for an eternity if it chooses to. Some of these state regulators are willing to reward their utilities for taking steps to make our nation's electricity usage more efficient—but others are not, so you do not even have a national policy today for energy efficiency. You have a lot of gobbledygook coming out of Washington about the need for it, but that immediately runs up against 10, 12, 15 state regulatory commissions. They will not allow a financial return on an investment by a utility in all of that. That is insane, but that is our system.

I am more optimistic about that pay-off for an investor who invests in all of these two-dozen or so technologies than I am on whether this country will get its act together sufficiently to keep the lights on. That said, not that China is that much more brilliant than we are, but they know they need to go full throttle on renewable energy just to keep up.

That is where we are going—we are up a creek. We have road-blocked coal, and we have taken an eternity to get nuclear back. That announcement yesterday doesn't mean that nuclear will be operational in 2017 because you still have this left wing element out there which says no nuke is a good nuke.

But here's my point: whenever there is a crisis there is investment opportunity. That strikes me as an opportunity to invest in dozens and dozens of different companies, all of whose prospects look very bright. Yet we are stuck on Wall Street in a 1950's mentality, where it is oil, natural gas, or coal, but it is basically just the same old stuff, nothing ever changes with energy. I mean how can these guys be so out of place? You know, and all the best minds on Wall Street frankly have gone over to Europe. Now Asia is emerging as the alternative energy hedge fund capital.

CS: Wow that is amazing. Well this has been remarkable, as I knew it would be, Bill. Thanks so much.

For more information on this contributor, please visit:
http://2greenenergy.com/renewable-energy-facts-fantasies/.

RENEWABLE ENERGY—
ANOTHER VIEWPOINT ON
PICKING STOCKS

According to Jeff Siegel at Green Chip Stocks, "We're at the dawn of a new age in energy. The days of alternative and renewable energy quietly playing out as a niche market are coming quickly to an end." Jeff and I have had numerous conversations on the subject, and there is no doubt in my mind that he's dead-on right here. This is the interview that I conducted via phone to his office in Baltimore, MD.

Craig Shields: Let me begin by asking you where you see the renewable energy industry going generally. It looks to me that there are a couple of dozen technologies here, all competing for space in the overall pie chart of energy consumption. If you were a betting man, which ones do you think are going to win?

Jeff Siegel: You know, I don't think it's a particular technology. If you take that pie chart, and coal has a chunk of the pie, natural gas has a chunk of the pie, and oil has a chunk of the pie—and then you have renewables. Within that wedge for renewables I think you have any number of things that fit into it, but they all still make up that one wedge. There is no one type of renewable energy that's going to take the lead; it's going to be a combination of all that we have available to us now and in the future.

CS: OK, I guess that's one way of looking at it.

What do you see are the gating factors to the success of renewables generally? Why aren't we there now?

JS: That's a good question. There are a number of reasons. At the risk of playing the conspiratorial card, if you want to talk about transportation fuels, everything that we have today is designed around this oil infrastructure. You know, we have gas stations all over the country so you have this trillion dollar infrastructure already in place and you can't just walk into the room one day and say, "OK, I have this really great technology, it will be more efficient, it'll allow us to have a transportation fuel that's not going to be so polluting, it's going to help the economy, but the only problem is there's no infrastructure set-up for it." You can't just go up to the gas station and do a quick charge for an electric vehicle.

You know, back in the day when cars had just come out, there were battery electric vehicles that competed with internal combustion engines. There were vehicles back in the day that went 20 to 30 miles on a battery charge. But essentially, the ICE (internal combustion engine) won out.

As far as utility scale generation, coal is cheap, and one of the main reasons it's so cheap is that we never really figure into the equation the liquidation of our natural capital. We can brush aside mercury in the water or pollutants in the air. No one asked us to pay for that, so we think of it as cheap. But what if you really had to break it down? Say a coal-fired power plant operator had to pay not just for the coal, and the mining of coal, and the transportation of coal—and the liquidation of natural capital that happened along the way, whether it's mercury in the water or greenhouse gas emissions. And what about moving the coal? You move coal either by rail or by truck, so that requires oil—it's either diesel or gas that's going to move the truck. There are so many little components involved there—and if you really broke it down maybe coal isn't so cheap—we've never had to pay those prices. So, it's easy to say,

"Well, the economics of coal make sense." But technically they don't. But the problem is we don't pay the real price.

CS: That's precisely right. Internalizing the externalities, as it were.

JS: Right, absolutely. And the good thing is I think more people are starting to at least consider that. I've noticed, I would say five or six years ago when I brought that up, you could hear crickets chirping in the room. Now people are saying, "Oh yeah, I have read Natural Capitalism" or "I understand this." People are reading more about it and at least considering it as a valid argument—because it is a valid argument.

CS: Right. Well, let's look at this from a financial perspective for a second. I want to get into the core of who Jeff Siegel is and what's made him successful. So, your ability to analyze good and bad bets, as it were, within the green sector, is based on what? What analyses do you deploy, and what makes those more successful and more accurate than somebody else's?

JS: I don't know if they're more successful. Any time you analyze a company or a stock, I don't think it's really any different doing that for renewable energy than anything else. When I started out at Agora, I learned how to do this from some of the smartest guys in the business. I mean, I sat and watched and read their stuff and learned how to do it.

I know quite a few people who invest but don't really do any research on the company or the market in general, and I've never really understood that. Most of the people that are successful in this business take the time to do the research—not just read about things, but going to visit the companies. Go visit the competitors. See what's going on in the space.

You know, when we first started Green Chip, we were one of the first letters around to cover renewable energy. And since then there have been a great number of new ones coming out and launching their own letter. And some of them are successful and some aren't—but the ones that are successful look at these things on a company-by-company basis, or maybe they'll just look at policy, but they won't look at supply and demand.

There are so many things that you have to consider with any investment. When we look at renewables we treat it like anything else. I've long been a self-proclaimed environmentalist; I've long been a supporter of alternative energy. I believe in it. But when I look at it from an investment standpoint I ask, "OK, what do we have in front of us?" We have the rapidly diminishing supplies of all our fossil fuels, not just oil, but coal and natural gas. Even with new natural gas finds, you don't know what the production rates are going to be. So you have fossil fuel depletion, and environmental impacts which are real. I mean, we can sit and argue and debate stuff all day but in the end, if you can't drink the water and you can't breathe the air then you've got to make a change. And if people want a glimpse of what it would be without environmental regulation, go to China, and you'll see firsthand just how bad it can be.

And on top of all this, you have the economics of energy that are starting to be exposed for what they really are. And I think that kind of ties in with the depletion of fossil fuels. We can start to see that now with oil, and I believe we will see that in another 10 to 20 years with coal, people will realize that we don't really have a 250-year supply. The USGS (US Geological Survey) says that there is a 100-year supply; I would suggest that it's even less.

CS: OK, but how can you tell a good renewable energy investment from a bad one?

JS: If I look at a solar stock I'm going to look at the technology, I'm going to look at funding, I'm going to look at sales, I'm going to look at competition, I'm going to look at policy—some sorts of companies are funded so well that maybe they're not going to live and die by policy as opposed to some other ones.

I also look at where they're based. Except for First Solar, I really wouldn't touch a US solar company right now, because the Chinese simply manufacture solar faster and cheaper than anybody else. Now,

I'm not saying Chinese solar is better; I'm just saying, strictly from an investor's standpoint, that I have to go where I know who is going to deliver at the end of the quarter. And if I compare most European or US solar manufacturers to a Chinese solar manufacturer, just on a cost basis, it's a no-brainer.

You have to take all those different kinds of things and put them into the equation. When I first met with the Green Chip publisher, they asked about the difference between investing in renewable energy and investing in anything else—and really there is no difference. You just have to do your research.

And I really believe in going that extra mile, even though it can be a pain in the neck sometimes. If I have to go look at a geothermal power plant in Idaho, flying from Baltimore to Utah and then driving from Salt Lake City to Raft River is not a fun day. But I've been in this business a while, and one thing I've learned is that some of these companies will tell you anything—and that most of the time you have got to see it for yourself. I've talked to companies that had this great technology or that were building this power plant and then I go to visit and there's nothing there! It's "planned." You just can't be afraid to get your hands dirty.

CS: Wow, that's very interesting. You bring up a great point about the international versus domestic scene. Let's take those—one at a time, if we can. You know, I had the pleasure of meeting Matt Rogers, the assistant to (Energy Secretary) Stephen Chu, and I must say that I'm convinced that if there isn't something done in the public sector to create a level playing field, I don't think this is ever going to happen. What do you think has been the role, and will continue to be the role, of the public sector in stimulating this migration to renewables?

JS: It's huge. I agree with you 100%. You know, without a level playing field we can't compete; that's just a fact.

Take a look at all the subsidies that go into fossil fuels. There is an environmental law research group that put together a list of all the subsidies from 2002 to 2008 for fossil fuels versus renewables. Fossil fuels got something like three times as much as renewables—and half of the renewable subsidy went towards corn-based ethanol.

You have two choices. You can either support renewables domestically, get them the support they need so that they can compete on a level playing field, or you can pull all subsidies for everything and then you have a real level playing field. But if you do that and everything comes to a crashing halt because all of a sudden you're price of food just went up by a factor of 10. We rely on 3000-mile tomatoes.

Without policy support nothing is going to change. I'm a big believer in a free market, a real, true, honest, free market—but I don't think we've ever really had that, at least as far as where it comes to energy as long as I've been alive.

To give you an example of this, it makes me crazy when the Chinese were going to come to Texas and going to build a wind farm. That was after how many opportunities that we had here in the United States to do that? We couldn't form the capital for any number of reasons. So the Chinese said, "OK, we have the money. We'll come and we'll set up a wind farm in your country and we'll profit from it. And then we said, "Well, wait a minute. You want stimulus money; we're not going to give you any because you're going to make that stuff in China." "Okay, no problem," they say. "We'll set up a manufacturing facility right here."

Now I'm certainly happy that we are putting people to work. I mean, at this point we need jobs so if someone's going to employ 300 workers in Texas to build wind farms, I don't care who's going to do it just make it happen. But at the same time, it kills me to know that because a level playing field does not exist, we would have to outsource not only our workers, but the actual facility as well.

CS: Right. This was Chuck Schumer trying to put a spear through this whole thing just because the Chinese were involved in it financially. That's unbelievable. We really love to cut off our nose to spite our face.

JS: Yes, absolutely. And then try to explain this to the people that live there that are still collecting unemployment that would kill to have a job, you know?

CS: Right. Okay, let me ask you a little bit about your bread and butter—financial newsletters. Who has an appetite for this? Are they individuals? Are they investing their own money? Are they investing other people's money?

JS: Well, it's primarily individual investors that buy these newsletters. I know that when I first got into this business, I didn't understand it at all. Do people buy newsletters for this stuff? My attitude has always been that if I want to invest I'll do the research and do it on my own. But I get it now; it's a legitimate industry where a lot of people invest a lot of money and they don't have time to spend their every waking second flying off to check out a wind farm or whatever it is they want to invest in.

I've always referred to Green Chip as an independent investment research service. And I do that because I want to distinguish ourselves from a lot of the other letters. It's a tricky industry because, you know, there are a lot of newsletters out there, and I'm not going to name names or anything, but there are lot of newsletters out there that in my opinion are dishonest. A lot of newsletters are "pay to play," where people will come knocking on the door saying "Hey, I want you to cover my company. In return, I'll give you so many shares of cheap paper." It's very common. I don't want to say the whole industry is like this, but there are certainly people out there that are like that, and it makes it hard for legitimate newsletter writers, because people want to lump you in that category.

I've had plenty of people come up to me at conferences and ask, "Listen, what would it take to get my company covered? How much is it going to cost me?" And my response is always the same: we don't get compensated for coverage. People rely on us to do the best job that we can to give them the best investment research that we can provide. And if you start getting into situations where people pay you, you have no objectivity, and people put their faith in something that's not real. So I'm not going to have anything to do with that.

The second reason is that I've long been a renewable energy advocate; it's something that's very close to me. If I weren't working in the newsletter industry, I would be doing something else in this space; it's a love of mine; it's a passion. And I would never want to dirty that up by being unethical. And I think that's one of the reasons that we've been really successful. A lot of these newsletters go out of business in a couple of years. We've been around six years now. And I think the reason is people have a lot of faith in us.

CS: Well, good for you Jeff. This has been great. Keep up the good work.

JS: Talk to you again soon.

For more information on this contributor, please visit:
http://2greenenergy.com/renewable-energy-facts-fantasies/.

CONCLUSION

I hope this book has been successful in communicating a few of the most important issues facing us as we struggle to replace environmentally unsound forms of energy with renewables, and that I've presented a few fresh and important viewpoints on this subject. In the course of this project, I've tried to "stay out of the way" and let the subject-matter experts do the talking. Yet, if the reader has stayed with me to this point, I can only assume he has some interest in my own summary of the subject.

Here are a few of my beliefs:

Tough Realities

Probably the most obvious result of these interviews was the re-enforcement of an idea that we probably should have known all along: We live in a world of tough realities, where an elegant solution simply does not exist.

This is why I took exception to an edition of David Brancaccio's NOW earlier in the year, for its gross oversimplification of the migration to renewables. In an attempt to inflame the viewer about the dangers of fracking (hydraulic fracturing, injecting water and chemicals deep underground to pry out gas locked away in tight spaces), the show told its viewers flatly, "We have renewable energy technology right now."

At a certain level, this, of course, is true; there are a dozen or so clean energy technologies that are quite functional. But without context, this statement is horribly misleading. Sure we have the technology now, but there are hundreds of issues that many thousands of people are diligently working on — that will ultimately enable renewables to be deployed in an economically, legally, and ecologically sound way.

Peak Oil

One thing you can say about Peak Oil is that there is precious little consensus on the subject. Personally, I don't understand all the attention the world pays to this concept. Is our path of extracting, refining, and burning crude oil sustainable? Of course not. But what is the significance of establishing the date—even to the decade—where we truly started—or will start—to decline in terms of oil production? Here are a few concepts that make this a point of academic interest only:

Oil prices are established by supply and demand, and, as we've seen over recent years, there is very little demand elasticity in this commodity—meaning that small changes in demand result in large fluctuations in price. These fluctuations also change the viability of certain extraction techniques. What makes no sense for oil at $50 a barrel is a terrific idea if oil's at $150.

The world's demand for oil, right now, is accelerating—but at a rate that no one can predict. Yes, we're on a course to have twice as many cars on the road by 2030 as we do now. But it's by no means a certain course. And how many of them will be electric or other alternative fuel? As we've seen in this one short book alone, the numbers vary all over the place.

What will happen politically and economically over the next decade or so—that will forward—or impede—the deployment of renewable energy? This will have an even bigger effect on oil supply and demand.

Technologically, what R&D-related improvements will happen with the forms of the renewables that have the most promise to move the needle rapidly, for example, solar thermal with molten salt storage.

For that matter (not that I'm rooting for them) but what improvements will we see in fossil fuel extraction techniques?

I'm reminded of the business plans I read in which some idiot has predicted his sales volume for 2016 out to three decimal places. Is this a joke? Can anyone be expected to believe that we have such a grasp of reality here in 2010—let alone a vision of the future—to make that kind of prediction?

Almost everyone who has carefully looked into this matter agrees that, even if there were no environmental issues at all, that the world cannot process oil (whether most of it is a fossil fuel or not) in sufficient supply as the demand for energy continues to grow. And, though I agree with Mr. Simmons that the consequences are dire, I find him unnecessarily pessimistic about the trajectory for renewable energy and electric transportation.

National Security

It's quite obvious that we fight wars over oil; I don't know how—or even why—anyone would attempt to deny that. Oil is a resource that lies at the very heart of our culture; wouldn't anyone be rightfully shocked if the US didn't fight to preserve its very life's blood? After all, it's not like we're exactly pacifistic as a nation. Can you think of something more serious to fight over?

Having said that, it seems to me that Mr. Woolsey is correct about almost everything else he said. In particular, there are huge national security issues militating towards an immediate migration away from oil. Yet he appears to see nothing sinister about the fact that, knowing all this, we're still borrowing a billion dollars a day to buy oil, pushing our country further into debt and empowering the world's truly evil interests. According to Warren Buffet, the total annual hemorrhaging for foreign oil to our national economy is seven hundred billion dollars. I'm not sure how anyone could conclude that there isn't some extremely sinister force behind this.

Externalities

I favor "internalizing the externalities," or forcing the producers and consumers of all forms of energy to pay the true and complete costs: increased healthcare, long-term environmental damage, etc. If this were to happen, renewable energy—even in its most extravagantly expensive forms—would be immediately perceived as the bargain of the century. Oil, coal, and nuclear businesses would be gone instantaneously. In the course of this project, I found that I'm not the only one with this idea; in fact, it seems to be gaining some level of traction.

Global Climate Change

Regarding global climate change, I have no ax to grind on the subject. But I don't mind going on record as saying that, based on numerous conversations with people like Dr. Ramanathan, I do believe in the theory. The fact that the world seems to have back-pedaled on this understanding is, in my estimation, largely a simple function of the propaganda on the subject. The oil companies have admitted to establishing sham research organizations to promulgate anti-global warming pseudo-science (though they say they've stopped).

Of course, I know that "deniers" claim the precise opposite, i.e., that the source of the propaganda is the people who stand to profit from research or mediation of the global warming problem. Though I'm skeptical of this, my only response is that, on a practical level, it really doesn't matter. As presented at numerous points throughout the book, there are many reasons to cut our consumption of fossil fuels that have nothing whatsoever to do with global climate change. I know there are people who don't believe the theory. But are there those who don't believe in cancer? Terrorism?

I know I'm not the first to point out something else about deniers' position, i.e., motive. They claim that global warming researchers may color their findings in order to encourage the flow of grant funding for their work—and it's undeniable that such a motive exists. But what would you guess represents more money (and thus more incentive to bias one's findings): the business of atmospheric research, or the business of selling of trillions of gallons of gasoline?

In any case, it will be interesting to see what happens here. I predict that it will become increasingly clear that the energy companies have acted very poorly in creating and covering up an enormous world health hazard. I forecast that in 50 years, these companies will be subjected to the same disdain that the tobacco industry is receiving today in terms of class-action lawsuits and broad societal condemnation.

Of course, the analogy falls down to the degree that cigarettes are a (deadly) luxury item (where energy is a necessity), and clean energy is still not viable in adequate supply at acceptable prices. But I predict that, over the coming decade, you'll see all manner of smoking guns around deliberately withheld information related to fossil fuels and public health. We have started to see this already, by the way, with pieces like 60 Minutes' treatment of coal ash in late 2009. (This was the quintessential 60 Minutes hatchet job — but it's a good sample of the scorn that's coming down the pike — both fair and unfair.)

Role of Government

I'm at sea about the role I'd like the public sector to play in this area. Pretty much wherever I look in government I see blatant corruption—and where I don't see corruption per se, I see a political process that is completely out of sync with the needs of our society. This, I think, is one of the central themes that emerged from many of the book's interviews: it is utter foolishness to expect politicians whose jobs are dependent on a two-year election cycle and big money donations to take actions where the value will not become evident until decades hence and run contrary to the greed of their paymasters.

But this is not to say that the public sector has no legitimate role in moving along the agenda of clean energy. The activities of organizations like NREL and the City of Santa Monica underscore this basic truth.

Most Promising Technology—Solar Thermal

In my opinion, solar thermal (with molten salt energy storage and high-voltage DC power transmission) should ultimately win the day in terms of the best, least expensive, most scaleable, safest energy solution. (I say "should" since, again, this is a matter of money and political power as well as technology.)

Making real progress toward an elegant solution to our energy woes in the US will require a federal government that exerts force under our crystal-clear eminent domain legislation to create corridors for transmission lines. I point out that we have created national pathways for the transportation of automobiles, railway cars, natural gas, etc.; there should be nothing new or scary about this. The fact that there is pushback on this, again, is simply the result of political posturing of those trying to delay the implementation of renewables so that they can continue to profit from existing technologies.

Readers will note that I'm anxious—perhaps over-anxious—to get to the point where two or three technologies have risen to a level of clear superiority, and the rest can be abandoned as superfluous. I admit that this process, if it occurs at all—will most assuredly take decades—but I think it's worth thinking about nonetheless. For instance, the day we're no longer constrained by delivering liquid fuels to our cars and trucks, the idea of biofuels will become far less exciting than it is today. Conversely, when we no longer need to generate electricity locally, technologies like solar thermal will be much more appealing.

Hydrogen

With all due respect to my colleague Steve Ellis, I side with those who are betting against hydrogen. The primary issue here, of course, is infrastructure. Even if all the stars align for abundant clean cheap ways to generate energy (that charges batteries and reforms methane or electrolyzes water into hydrogen), we are still faced with an incredible challenge to create a brand-new infrastructure to deliver fuel across a country of 3.5 million square miles.

Cold Fusion

I believe that cold fusion exists, and that there is a trajectory under which it could be used to generate clean energy. I will be very surprised if it happens, however, as I believe that we will have solved the problem in other ways before we get the research money and the many decades that will be required. (See my comment above on solar thermal.)

Driving Habits and Social Paradigms

Not that I know the first thing about sociology, but I do not agree with Dr. Kearl. In fact, to be honest, I was completely stunned with what he said. I'm predicting a paradigm shift in the way people view their cars and the activity of driving in which people begin to disassociate themselves from the brand (and style and size) of the car they drive.

Having said that, it would not surprise me to see this shift initiate in some other part of the world that has a deeper social consciousness than the place in which Dr. Kearl makes his home. (In fact, during the interview, he said, "You *do* realize that I'm in Texas, don't you?") He has a point. I hope the fine people of Texas will not take offense, but we must acknowledge that they live in the Big Oil and death penalty capital of the world; I have to think that there are places that are more likely to spawn a revolution in social enlightenment.

New Energy—A Huge Industry

From a financial perspective, I don't think there is any doubt that clean energy and transportation—and all the other ingredients that go into sustainable living—represent an incredible bonanza. By the time this book is published, 2GreenEnergy readers will notice a series of reports and newsletters available on the site, assembled by some of the top financial minds on Earth. We hope to be able to provide readers sound advice and direction across a wide gamut of industry sub-categories.

Let me conclude by reminding readers what I've been telling my friends about my interest in the subject: it's part science, but to a large degree it's big business and big politics. I'll end with an excerpt from a conversation on the 2GreenEnergy blog:

I happened to be writing about nuclear power one day and noted: I believe that there is no future whatsoever for the industry. I know there are people who disagree (and that I'll be hearing from several of them any minute). But to me, no amount of money and the lobbying, the subterfuge, and the disinformation that money buys will get the nuclear people past the incredible dangers, outrageous costs overruns, and decade-long delays that are intrinsic to the very nature of what they do.

With all their financial (and thus political) strength, I don't doubt that you'll continue to hear claptrap about supporting nuclear. There is a word for this: corruption; it's a regrettable but deeply entrenched part of our lives — whether we recognize it or not. But having said this, I very much doubt that you'll live to see another new nuke actually put into operation in the US.

Blogger "Hank" replied: … Corruption, subterfuge, and disinformation. I hate these, but I must agree with you that those are the biggest problems of our existence on this planet. How do we overcome basic human nature: greed, selfishness, lies, etc?

I answered: Thanks for writing, Hank. You ask the same question that I constantly pose to myself. And, to be honest, my answer differs according to my mood. Often I think there is no hope here; I see a dwindling spiral brought on by ruthlessness and greed, fueled by society's tendency toward apathy and numbness. Other days I wake up believing that we can somehow turn this around. I can tell you one thing: If we do revert this trend, it will be because people like you and I spoke up. Thanks again for your thoughts.

And so, in closing, I ask you: please speak up.

Craig Shields lives with his wife Becky, his children Jake and Valerie, and four dogs near Santa Barbara, California.

EPILOGUE—THE GULF OIL SPILL AND ITS IMPACT ON RENEWABLES

Many of my friends have asked my opinion on the likely impact of the Gulf oil spill on the trajectory for renewable energy. And although one might think that I'd be in a reasonably good position to answer a question like that directly and accurately, in truth, it really is hard for me -- or anyone, I believe — to predict the effect of this catastrophe on the world's energy policy going forward. I offer a few points for discussion:

Many people suggest that, as horrific as the spill is, it comes with a "silver lining," i.e., accelerating the demand for a replacement for oil as our predominant energy supply, brought about by an increased awareness of the many dangers of oil. Oh really? So the general public—normally fast asleep—has awakened? So a large flock of sheep had an epiphany on the dangers of oil and created a firestorm of outrage at the oil companies? So what? The same political forces that have continued to grant oil companies enormous subsidies through the last half century and made gasoline/diesel 98+% of our transportation fuel—even when we became aware of the dangers many decades ago — are still in place. And now those forces are working harder than ever. Do you think the corporate powers and (by far) the biggest lobby on the planet are updating their résumés and looking for new careers because of a lousy oil spill?

In addition to the big politics and big money, there legitimately are technology issues. Of course, these issues would have been largely mitigated, or eliminated entirely, if we had done what we should have been doing since the oil embargoes of the 1970s: running 1000 miles per hour toward electric transportation and various forms of renewables. Now, our oil addiction is so severe that the consequences of moving away from it are, like withdrawing from any addiction, quite unpleasant.

And consider global climate change. Some people say that the oil spill negates any point that the "deniers" may have had — i.e., now the validity of the global climate change theory no longer matters. Of course, that's been the case for a long time as well. If you're looking for a reason to break our oil addiction, the argument about global warming has been moot for many years; it's been obvious to most of us that there are five or six different equally compelling reasons. I know there are people who disbelieve the climate change theory; I run into them all the time. But are there people who don't believe in terrorism? In the consequences of a ballooning national debt? In lung cancer? In the dangers of weak national security? In ocean acidification? The spill is certain to weaken the position of the oil zealots (and whatever forces control them) — who try so hard to sell us on the idea that "oil business as usual" is a reasonable path towards a sustainable civilization.

So I suppose that there really is a silver lining here. It is precisely that now, anyone and everyone (you don't have to be a clean energy editor/business analyst) can see the truth for what it is. There is one and only one winner in oil, namely the oil companies themselves. Recall the tobacco companies of the 20th Century, and their product — the only legal one that when used as directed causes death. At a certain point we all realized that cigarettes were very good for Philip Morris investors and executives—but that they were very bad for literally everyone else on the Earth. The issue is the same here. The oil companies are the sole beneficiary of oil. And now, finally, it's clear to everyone.

Let's acknowledge that we made a grievous mistake in the 1970s/1980s — and move on. And let's keep our eye on the ball this time. Dropping the ball once is not license to drop it again. Use this as a litmus test for our leaders: an elected official who is really on your side (if there actually is such a thing) will take whatever political risks may come his way to stay the course in the development of clean energy solutions.

But it's up to you and me to insist that our leaders do that. In case you haven't noticed, they don't do things because they're right; they do them because they're forced.

FOR MORE INFORMATION

I hope I've provided a bit of insight into the renewable energy industry, and I'd like to close with an invitation. Please become a part of the 2GreenEnergy community. In particular, if you go to: http://2GreenEnergy.com/renewable-energy-facts-fantasies, you'll notice an ever-growing set of content from the book's 25 contributors. Over time, you'll see articles, reports, bios, videos, and podcasts—all aimed at making sense of this multi-trillion dollar industry.

Also, please enjoy—and contribute to—the blog at http://2GreenEnergy.com/blog. I will most certainly look forward to hearing from you.

Best regards,

Craig Shields
Editor
2GreenEnergy.com

APPENDIX

ALL THE PHYSICS THAT ALMOST EVERYONE NEEDS TO KNOW

Although some may object to my oversimplifications, most of the science behind renewable energy is high school stuff. In fact, most of it is rooted in the fact that our sun radiates enormous quantities of energy onto the Earth each day—and has been doing so for billions of years. Photovoltaics, solar thermal, wind, and hydrokinetics are all based on the idea that—even in our energy-hungry 21st century—Earth receives more than 6000 times more energy each day from the sun than the 6.8 billion of us consume. Even the energy in fossil fuels comes ultimately from the sun; the chemical energy in coal, petroleum, and natural gas is derived from dead animals and plankton, all courtesy of the sunshine that took its eight-minute-long journey through space to us hundreds of millions of years ago.

So if the science here is fairly easy, let's take a quick look at it. Here, I present a few principles of basic physics that are most relevant to the discussion of renewable energy. You may want to take out a piece of paper and a pencil, and draw examples of the concepts here; most people learn best when they can envision real examples of things. And the good thing about physics—at least as it was understood up until about 1900—is that it covers real, perceptible things. Fortunately, the science behind most of today's efforts to harvest, store, and distribute energy are rooted in science that has been well understood since the early 19th Century. While the theory of relativity—and (even worse) quantum physics—describe things that are not part of our day-to-day observable world and tend to be incredibly counter-intuitive, the concepts we are talking about here can easily be drawn.

Here are some concepts that we need ready to hand:

Energy—The ability to do useful work, energy comes in different forms that are forever transforming themselves into one another:

Potential—The capacity to do work, e.g., potential energy can be stored in a rock that has been lifted against the force of gravity.

Kinetic—The energy of motion, e.g., that rock as it falls back to Earth.

Chemical—The energy stored in bonds that hold molecules together, e.g., the wax in a candle.

Heat—The transfer of energy associated with high-temperature masses, e.g., burning candles.

Nuclear—Energy released by the breaking apart or fusion of parts of an atom, according to Einstein's theory of special relativity, with its famous equation stating the equivalence of mass and energy: $E=MC^2$.

Electrical—The energy associated with electrons moving through a conductor.

Electromagnetic—The energy associated with electrical and magnetic fields that oscillate back and forth with a spectrum of different frequencies.

Here are some examples of common processes in which energy is converted from one form to another:

When a candle is burned, the energy that was stored in the chemical bonds in the wax is turned into the same amount of energy in the form of light and heat.

When you fill a balloon, the muscles surrounding your lungs convert some of the chemical energy from the food you've eaten into mechanical energy that compresses the air in the balloon. The balloon now has potential energy that can be converted into yet a different form, kinetic energy. I.e., if you release the balloon, it will zip around the room.

Conservation of Energy

The law of conservation of energy says that energy is never created or destroyed; it simply changes forms. Thus, no device exists—or could possibly exist—that delivers more energy than is put into it.

Now obviously this is a controversial thing to say. There will never be a process that violates the conservation of energy? Never is a long time! Perhaps it's safer to say that the conservation of energy is a paradigm, and that it's hard for serious scientists of the 21st century to consider that it will ever be replaced.

Let's examine the concept of power, which is the amount of energy converted from one form to another in a unit of time. A 100-watt lightbulb burning for one hour consumes 100 watt-hours of electric energy (some of which will be converted to light and some to heat). Do not confuse power with energy. Power (often measured in Watts) is the rate at which energy is being converted. Energy (often measured in Watt-hours) is the result of power being applied for a certain amount of time.

Take a minute or two and think of some more day-to-day occurrences in your life that involve the conversion of energy from one form into another. Here's one: the chemical energy that existed in the food you've eaten is released and made available to your body for a multitude of purposes, converted into the following forms (which, if you added them all up, would equal the original chemical energy in the food before you digested it):

You store some of that chemical energy as fat in the tissues you have for this purpose, for use at some later time.

You walk up a flight of stairs, applying a force to overcome the force of gravity that is resisting you, across a distance. You gain potential energy equal to the weight of your body times the height of the staircase.

Your muscles heat up, and some of that heat energy is dissipated through your skin. Your body perspires to cool itself, knowing that the presence of this perspiration causes evaporation which removes heat energy from your body turning that liquid to a gas.

You speak, converting some energy into pressure waves of the sound produced by your vocal chords.

Efficiency

Another concept central to our thinking here is efficiency. No process by which energy is converted to one form or another happens without some of the energy, regardless of how minute, being converted in a form that is lost, i.e., not converted to useful work. Here are some examples:

Solar panels convert sunlight to electricity, but the output of the solar panel is far less than 100% of the sun's energy that fell upon the panel. Some of that energy went into heat, or some other form of non-useful energy. As discussed earlier, the most common forms of solar panels are only about 20% efficient, meaning that about 80% of the incident sunlight is not converted to electricity.

Burning a gallon of gasoline in the engine of your car produces a great deal of wasted energy in the form of heat and sound, in addition to the (useful) kinetic energy that it delivers to the wheels of your car. As explained in this diagram published by the federal government, in most cases, less than 15% of the chemical energy of gasoline is converted to useful work.

Once we wrap our wits around these basic ideas, we're in a terrific position to understand most discussions of energy—renewable and otherwise.

Here are two examples to make this clear.

Hydrokinetics

Every day, the energy from the sun evaporates some liquid water into steam, some of which is condensed into clouds, the precipitation from which forms a river, some of which begin in high altitudes. The kinetic energy of the water flowing back downhill can be converted into electrical energy. But the conservation of energy tells us that the most electricity one can possibly hope to generate from this water is equal to the potential energy it had before it started to flow (which is the weight of the water times the height of the elevation from which it fell). It is for this reason that hydrokinetics cannot provide more than a certain amount to the overall energy picture, regardless of how many dams, how efficient the turbines, etc.

Solar

As mentioned above, the Earth receives 6000 times more energy from the sun every day than mankind currently uses for all its purposes: transportation, heating, air conditioning, etc. Put another way, if we had the capability of capturing and distributing 1/6000th of the sun's energy, we would not need to burn another lump of coal, spilt another atom, or distill another ounce of gasoline.

Electricity and Magnetism

In certain circumstances, charged particles (electrons) become detached from their atoms and are forced to flow in a certain direction, providing the capability to perform useful work, not unlike the flow of water downhill in a river. Electrical energy is generated by converting mechanical energy—forcing an electrical conductor to move through a magnetic field.

Here are a few concepts:

Current—The number of electrons passing a certain point in a given unit of time, normally measured in amperes.

Voltage—The amount of force that is exerted on the charged particles, tending to cause a current to flow, normally measured in volts.

Resistance—The opposition to a flow of current, normally measured in ohms.

Ohm's Law—The more voltage and the less resistance, the more current flows. This is expressed in the formula V=IR, or voltage equals current times resistance. Notice how intuitive this is. If you had water flowing through a hose, the more force (voltage) with which you pushed the water, the more water would flow (current). The more restrictive the hose (resistance) the less water would flow.

Battery—A device that uses chemical reactions to store and deliver electrical energy.

Capacitor—A device that uses electrical conductors that are physically separated from one another to store and deliver electrical energy.

Direct current—Movement of electrons in one direction, the type of current produced by a battery.

Alternating current—Movement of electrons back and forth.

Inverter—A device that converts direct current to alternating current.

Electrical power—As above, power is the amount of energy converted from one form to another in a given unit of time. The power delivered by electricity (measured in watts) equals the voltage (volts) times the current (amps).

I encourage readers to analyze all ideas about energy—renewable or otherwise—to this discussion. When someone says, "This car runs on water," ask yourself: water? Water is really "burned" hydrogen and oxygen, i.e., the result after these two elements release energy by joining together. It's rather like saying, "Let's build a fire using those ashes for fuel." Sorry, they've already been burned; the chemical energy that was once stored in the carbon bonds of the wood has already been converted into the heat (and small amounts of light and sound) of a fire. What you have there is the result of that process.

I got an email from a friend announcing a "miracle" car that runs on air. Actually, it runs on compressed air. The energy required to compress the air is stored in a tank and converted into kinetic energy. But trust me, there is not one bit of energy delivered to that car's wheels that didn't go into compressing the air in the first place.

Again, once this is all understood, we're in a much better place from which to comprehend the energy industry.

INDEX